Esra Russell, Oktay K. Pashaev

Oscillatory Models in General Relativity

De Gruyter Studies in Mathematical Physics

Volume 41

Esra Russell, Oktay K. Pashaev

Oscillatory Models in General Relativity

DE GRUYTER

Physics and Astronomy Classification Scheme 2010
98.80.Jk, 95.30.Sf, 02.30.Gp

Authors
Dr. Esra Russell
New York University Abu Dhabi
Saadiyat Island
Division of Science
129188 Abu Dhabi
United Arab Emirates

Prof. Dr. Oktay K. Pashaev
Izmir Institute of Technology
Gulbahce Campus
Department of Mathematics
35430 Urla-Izmir
Turkey

ISBN 978-3-11-051495-7
e-ISBN (PDF) 978-3-11-051536-7
e-ISBN (EPUB) 978-3-11-051522-0
Set-ISBN 978-3-11-051537-4
ISSN 2194-3532

Library of Congress Cataloging-in-Publication Data
A CIP catalog record for this book has been applied for at the Library of Congress.

Bibliographic information published by the Deutsche Nationalbibliothek
The Deutsche Nationalbibliothek lists this publication in the Deutsche Nationalbibliografie; detailed bibliographic data are available on the Internet at http://dnb.dnb.de.

Typesetting: le-tex publishing services GmbH, Leipzig
Printing and binding: CPI books GmbH, Leck
♾ Printed on acid-free paper
Printed in Germany

www.degruyter.com

Contents

Introduction

> We are just an advanced breed of monkeys on a minor planet of a very average star.
> But we can understand the Universe. That makes us something very special.
>
> |Stephen Hawking|

Modern mathematical cosmology was constructed between 1907 and 1915 by Albert Einstein, in which he used his gravity model to understand the dynamics of the universe. This model was built using his general theory of relativity (also known as general relativity), which was constructed in 1916 using Riemannian geometry. Although his model stated that the universe is expanding, observations did not support this prediction until 1922. In 1922, Alexander Friedman used the modified equations of general relativity to obtain the same result as Einstein of an expanding universe. Since there was no observational evidence of cosmic expansion, Einstein modified the field equations of general relativity by adding a term called the cosmological constant. The cosmological constant provides a repulsion to compensate the gravity attraction and to stop expansion, leading to a static model. In 1929, observational evidence changed the fate of the Einstein's general relativity model: Edwin Hubble's research on the red shift of distant galaxies confirmed the prediction that the universe is expanding. As a result, Einstein considered the cosmological constant as his biggest blunder. Following Hubble's discovery, cosmologists started to construct expanding universe models in the context of general relativity, in which the consequences of different assumptions about the distribution of matter in the universe are investigated. Therefore, the initially simple cosmological models have been replaced by more complex models taking into account nonlinearity and dissipation.

The modern cosmological models are based on the Friedman–Lemaitre family of models, which are built from the Robertson–Walker (1934) spatially homogeneous and isotropic geometries. Although there is observational evidence supporting these models on the largest scales, at smaller scales they do not provide a good description. The questions are [19]: On what scales is the geometry of the universe nearly Friedman–Robertson–Walker (FRW)? Why is it FRW? How did the universe come to have such an improbable geometry?

The answer to these questions can be found in inflation theory [22, 31]. According to this theory, the quantum fluctuations in the very early universe formed the seeds of inhomogeneities that could then grow. To examine these questions one needs to consider the family of cosmological solutions in the full state space of solutions, allowing one to see how realistic models are, related to each other and to higher symmetry models including, in particular, the Friedman–Lemaitre models.

Here we discuss general techniques for examining the FRW-type family of models and their generalizations, which could be useful in describing the universe at large scales. First of all, in FRW-type approaches the universe is characterized by cosmic-scale parameters, which are functions of the global time variable. From this point of

https://doi.org/10.1515/9783110515367-201

view the isotropic universe is a dynamical system with one degree of freedom. But at smaller scales the anisotropy of the universe could be important, which is why one can consider a more general situation, with three different scale parameters depending on one global time. In this case, which is in the class of the so-called Bianchi family of universes, we have a dynamical system with three degrees of freedom and the FRW universe appears as the only symmetric reduction valid for the isotropic case. The general anisotropic case can be described by the Riccati equation and this equation admits transformation to the time dependent damped harmonic oscillator. This is why such models are called oscillatory models of the universe.

A time dependent metric also leads to the problem of the complexity of the physics in a time dependent background. Brandenberger [9] showed that inflationary metrics also imply time dependent frequency for the gravitational wave modes. This allows us to extend the canonical quantization method for nonunitary time evolution to include the quantization formalism for a parametric oscillator. [29] studied a harmonic oscillator with a time dependent frequency and a constant mass in an expanding universe. In the inflating case, the FRW metrics produce a damped harmonic oscillator equation for the partial waves of the field $h_{\mu\nu}$ [21]. [1] discussed the canonical quantization of nonunitary time evolution in an inflating universe. They considered gravitational wave modes in the FRW metrics in a de Sitter phase, then applied the quantization method to the damped oscillator mentioned above. Following this, the doubling of the $h_{\mu\nu}$ partial waves, which was called double universe, was shown by [3].

If damped oscillatory models are very important at small scales, the natural question is what happens upon quantization of these models, when one cannot neglect quantum fluctuations. One of the first approaches to quantize the damped harmonic oscillator is to start with the classical equation of motion, then find the Lagrangian and then the Hamiltonian, which will lead to the Hamilton equations of motion, and finally to quantize them by the canonical formalism method. This approach is called the (Bateman)–Caldirola–Kanai model, which derives quantum mechanics from a dissipative Hamiltonian. This Hamiltonian was actually proposed earlier by Bateman, but in a classical context [5]. This approach has the attractive quality of providing an exact solution, in essence because the classical equation of motion has an exact solution and formal quantization merely has the effect of converting the classical variables into operators. A second approach uses an interaction Hamiltonian and applies perturbation theory. One is a rather simple system (the undamped harmonic oscillator) that we construct, but an environment of the damped harmonic oscillator also exists. These, in fact, close the system, which creates a realistic or artificial embedding within a larger system that preserves energy. This way, Hamiltonians that describe a total, conserved energy can be obtained. An example of this line of thought is the so-called doubling the degrees of freedom approach. In fact, this idea can also be traced back to a Hamiltonian that was coined by Bateman, the so-called dual Hamiltonian [5]. The idea is that the damped oscillator is coupled to its time reversed image oscillator, which absorbs the energy lost so that the energy of the whole system is conserved or closed. In fact, since the phase space of the whole system describes the damped harmonic oscillator and its image, the degrees of freedom are effectively doubled. Another way of looking at this is that adding a time reversed oscillator restores the breaking of the time reversal symmetry. Difficulties arose during earlier attempts to elaborate this idea, such as time evolution leading out of the Hilbert space of states, but later a satisfactory quantization could be achieved within the framework of quantum field theory [8, 14]. The doubling of degrees of freedom approach has the conceptual disadvantage that the environment to which the damped harmonic oscillator is coupled is artificial. However, the word 'artificial' is only used for well-known systems. Since the structure of universe is not well defined, this approach has the advantage of showing the main form of dissipation as a system.

Apart from these approaches for a damped harmonic oscillator with a constant frequency and damping coefficient, the general form of a time dependent Hamiltonian that describes a classical forced oscillator with a time dependent damping coefficient and frequency were studied by [23]. This kind of system was also considered by other studies [27, 33]. Moreover, Kim demonstrated that canonical transformations in classical mechanics correspond to unitary transformations in quantum mechanics [43]. Additionally, Kim and Lee studied time dependent harmonic and anharmonic oscillators and found the exact Fock space and density operator for a time dependent anharmonic oscillator [28].

The goal of the first part of this book is to study the dissipative geometry of universe models in general relativity in the following contexts.

In Chapter 1, the fundamental definitions of general relativity, such as Christoffel symbols, Riemann tensor, Ricci tensor and Ricci scalar (Section 1.1) as well as the definitions of the Einstein field equations both in the presence and in the absence of matter, and the definitions of the cosmological constant, are given. We also discuss one of the most important tensors of general relativity: the energy momentum tensor, which tells us the energy-like aspects of the system. In Section 1.2, we discuss the universe as a dynamical system based on time dependent and scale factor dependent metrics. In this framework, the solution of the Einstein field equations and the derivation of the equation of state are given. These equations are particularly important since they can help us understand the universe as a dynamical system.

In Chapter 2, the construction of the universe models begins with the idea that the universe on large scales is isotropic and homogeneous. As a result, the Friedman universe models are considered, including four basic group of models: static, empty, non-empty with zero cosmological constant, and non-empty models with nonzero cosmological constant (Section 2.1). In Section 2.2, Milne's model and its fundamental properties are discussed, and Milne's model and the Friedman models are compared.

In Chapter 3, anisotropic and homogeneous universe models are investigated in terms of different density and pressure functions. In Section 3.1, the general solution of the field equations is obtained with respect to the anisotropic and homogeneous metrics. In the subsequent subsections, the particular solution of the field equations in the radiation dominated model is given.

In Chapter 4, the linearization of the Einstein equations are given, which produces gravitational waves on the Minkowski background and from the Fourier expansion of the field. The Fourier component of the field satisfying the harmonic oscillator equation with constant frequency (Section 4.1) is obtained. Following this, in Section 4.2, the linearization of the same equation on the de Sitter background produces damped harmonic oscillator systems with respect to the Bateman approach; the double universe models can be formed with respect to this approach. In Chapter 5, applying the factorization procedure to Friedman equations, bosonic and fermionic models of FRW universe are described. In Chapter 6, we consider an oscillatory universe models with time dependent gravitational and cosmological constants.

The second part of this book is devoted to study of the variational formulation of time dependent harmonic oscillators.

The Lagrangian and the Hamiltonian descriptions are crucial to understand the damped oscillator in quantum and classical theory. Hence, the background of these descriptions is given in Chapter 7. In Section 7.1, we give the definitions of the generalized coordinates and the velocities. In Section 7.2, a formulation for the study of a mechanical system, which is called least action principle, is discussed. In Section 7.3, the Hamiltonian and Hamilton's equations, the Poisson brackets and the properties of the Poisson brackets are discussed. In Section 8.1, the solution of damped harmonic oscillator is considered for three different cases: overdamping, critical damping and underdamping. An extension of analytical mechanics to include dissipation, is discussed in Section 8.2. In Section 8.3, we give the definition of the Bateman dual description, and using this approach we investigate the Lagrangian and the Hamiltonian functions for doublet damped oscillator systems. In Section 8.4, the time dependent Hamiltonian with time dependent mass satisfying the standard damped harmonic oscillator equation, which is called the Caldirola–Kanai Hamiltonian, is given. In Section 8.5, the Caldirola–Kanai Hamiltonian with a constant damping coefficient and frequency is quantized.

In Chapter 9, the two different formulations of damped oscillator with time dependent damping and frequency are related to the self-adjoint extension of the Sturm Liouville problem (Section 9.1) are given. In Section 9.2, the particular representations for the time dependent frequencies and the damping coefficient functions, related with different special functions are discussed. In Chapter 10, the Riccati representation of the special functions as oscillator-type problems is considered and some particular cases are given. In Chapter 11, quantization of damped oscillators with time dependent damping and frequency is considered. Exact quantum solution of this problem in Gaussian form is constructed in terms of the Riccati equation and the classical damped parametric oscillator, studies in previous chapters.

Part I: **Dissipative geometry and general relativity theory**

1 Pseudo-Riemannian geometry and general relativity

The two theories established by Albert Einstein – special relativity and general relativity in 1905 and 1915 respectively – are the modern theories of space and time. These theories changed our view of Newton's concepts of absolute time.

In both special and general relativity theories, the notions of separate vectors in space and time are abandoned, and the notions of spacetime and four-dimensional quantities are introduced. In this four-dimensional spacetime, the separation between two events is given by the spacetime interval, also called the metric, ds^2. In special relativity, the spacetime interval in four dimensions is given by the Minkowski (four-dimensional, flat) metric,

$$ds^2 = \eta_{\mu\nu} dx^\mu dx^\nu \,, \tag{1.1}$$

where $\eta_{\mu\nu}$ is the metric tensor,

$$\eta_{\mu\nu} = \begin{pmatrix} -1 & 0 & 0 & 0 \\ 0 & 1 & 0 & 0 \\ 0 & 0 & 1 & 0 \\ 0 & 0 & 0 & 1 \end{pmatrix} , \tag{1.2}$$

and x^μ represent coordinates in the Minkowski space,

$$x^\mu = (ct, x, y, z) \,. \tag{1.3}$$

As can be seen, the Minkowski spacetime metric/interval is similar to the Euclidean space. For example, in Euclidean space, the infinitesimal spatial distance between two points is simply $ds^2 = dx^2 + dy^2 + dz^2$. The main difference is that while all the space coordinate contributions are positive, the time coordinate appears with negative sign in the Minkowski metric.

Additionally, events in general relativity occur in four-dimensional, curved spacetime rather than in flat Minkowski spacetime. Curved spacetime is defined by pseudo-Riemannian geometry, where the separation between two events like in the Riemannian spacetime is

$$ds^2 = g_{\mu\nu} dx^\mu dx^\nu \,, \tag{1.4}$$

in which $g_{\mu\nu}$ is called the Riemannian metric tensor. In contrast to the Minkowski metric $\eta_{\mu\nu}$, the pseudo-Riemannian metric is coordinate dependent, $g_{\mu\nu}(x)$. The metric components transform as a tensor

$$g_{\mu\nu} = \frac{\partial x^\alpha}{\partial x^\mu} \frac{\partial x^\beta}{\partial x^\nu} g_{\alpha\beta} \,, \tag{1.5}$$

where the partial derivatives $\frac{\partial x^\alpha}{\partial x^\mu}$ and $\frac{\partial x^\beta}{\partial x^\nu}$ form transformation matrices of the basis vectors.

https://doi.org/10.1515/9783110515367-001

In the next section, the definitions and formulations of general relativity theory will be given in a general framework.

1.1 Curvature of spacetime and Einstein field equations

Suppose a general coordinate system with general basis e_α and a general four-vector V, then this four-vector can be represented as

$$V = V^\alpha e_\alpha \,, \tag{1.6}$$

in which V^α are the vector components. The first derivative of this vector in terms of general coordinates becomes

$$\frac{\partial V}{\partial x^\beta} = \frac{\partial V^\alpha}{\partial x^\beta} e_\alpha + V^\alpha \frac{\partial e_\alpha}{\partial x^\beta} \,. \tag{1.7}$$

Here, the partial derivatives of the basis vectors are

$$\frac{\partial e_\alpha}{\partial x^\beta} = \Gamma^\mu_{\alpha\beta} e_\mu \,, \tag{1.8}$$

where $\Gamma^\mu_{\alpha\beta}$ stands for the Christoffel symbols. If we substitute the derivative of the basis vectors (1.8) into the first derivative of the four-vector, we obtain

$$\frac{\partial V}{\partial x^\beta} = \underbrace{\left(\frac{\partial V^\alpha}{\partial x^\beta} + V^\mu \Gamma^\alpha_{\mu\beta} \right)}_{components} \underbrace{e_\alpha}_{basis} \,. \tag{1.9}$$

The components of the first derivative of the four-vector give the covariant derivative

$$V^\alpha_{;\beta} = V^\alpha{}_{,\beta} + V^\mu \Gamma^\alpha_{\mu\beta} \,. \tag{1.10}$$

Note that covariant derivative of a vector is a tensor. As is shown, the covariant derivative specifies a derivative of a four-vector along tangent vectors in curved spacetime. Here the connection, or the Christoffel symbol $\Gamma^\alpha_{\mu\beta}$, holds some important properties:

(1) it is symmetric: $\Gamma^\alpha_{\mu\nu} = \Gamma^\alpha_{\nu\mu}$
(2) it is torsion-free (no twist of a moving frame), indicating that the metric is covariantly constant $g_{\mu\nu;\beta} = 0$.

The torsion-free property of the Christoffel symbols is particularly important. Similarly, the partial derivative of the metric tensor in special relativity is zero. However, in the arbitrary coordinate system of the pseudo-Riemannian geometry, the partial derivative of the metric tensor will not give zero, since the metric components are coordinate dependent. In the latter case, connections can be thought of as inertial forces.

Computationally it is very difficult to obtain the Christoffel symbols. However, for symmetric connection compatible with the metric, it is much easier to calculate the

Christoffel symbols directly from the metric tensor:

$$\Gamma^{\nu}_{\beta\mu} = \frac{1}{2} g^{\nu\lambda} \left(g_{\beta\lambda,\mu} + g_{\lambda\mu,\beta} - g_{\beta\mu,\lambda} \right) , \tag{1.11}$$

where the matrix $g^{\nu\lambda}$ is the inverse of the matrix $g_{\nu\lambda}$, defined as $g^{\lambda\mu} g_{\sigma\mu} = \delta^{\lambda}_{\sigma}$. Starting from this point, we will see that the metric plays a central role calculating many important characteristics of the curved spacetime.

To understand the complete theory of gravity, two important questions must be answered: How do particles behave in curved spacetime and how does matter curve spacetime?

1.1.1 Geodesics

The first question above is answered by postulating that free particles follow timelike and null geodesics. The geodesics are natural generalizations of straight lines in curved spacetime. In other words, a geodesic is the shortest path between two points on a curved manifold. Now we will demonstrate how to formulate a geodesic mathematically. To do this, one can make use of parallel transport. Following this idea, we take a vector $\vec{V}$ and parallel transport this vector along the curve. This simply means that we take the derivative of the vector $\vec{V}$ with respect to the parameter along that curve, which will be zero in a flat space or inertial coordinates,

$$\frac{d\vec{V}}{d\tau} = 0, \quad \vec{u} = \frac{dx}{d\tau} = 0 . \tag{1.12}$$

However, in a general coordinate system, the generalization of the same setting indicates that we take the covariant derivative of the vector $\vec{V}$ along the tangent vector of the curve, which is going to be zero:

$$D_{\vec{u}} \vec{V} = 0 \rightarrow V^{\alpha}{}_{,\beta} u^{\beta} + V^{\mu} \Gamma^{\alpha}_{\mu\beta} u^{\beta} = 0 . \tag{1.13}$$

We can easily see that in Minkowski spacetime the Γ term disappears above and the covariant derivative of $\vec{V}$ simply reduces to the classical case as in equation (1.12). Therefore, the geodesic is simply a curve, where a tangent vector is parallel transported. In this framework, we are allowed to change the vector components V to the components of the tangent vector u in equation (1.13). Therefore, the geodesic equation is obtained as

$$\frac{du^{\alpha}}{d\tau} + \Gamma^{\alpha}_{\mu\beta} u^{\mu} u^{\beta} = 0 . \tag{1.14}$$

Taking into account that the tangent vector $u = dx/d\tau$ is parametrized, the geodesic equation (1.14) can be rewritten in the following form:

$$\frac{d^2 x^{\alpha}}{d\tau^2} + \Gamma^{\alpha}_{\mu\beta} \frac{dx^{\mu}}{d\tau} \frac{dx^{\beta}}{d\tau} = 0 . \tag{1.15}$$

This geodesic equation represents the equation of motion not only in Minkowski (pseudo-Euclidean) spacetime, but also in the curved spacetime of general relativity theory.

1.1.2 Riemannian curvature tensor

Another important mathematical tool we can obtain using parallel transport is the Riemannian curvature tensor.

We parallel transport a vector, V^α, around a closed loop and then compute the change in the vector so that this change gives the formulation of the Riemannian curvature tensor. The Riemannian curvature tensor is the expression of the intrinsic curvature

$$R^\alpha_{\mu\beta\nu} = \Gamma^\alpha_{\mu\beta,\nu} - \Gamma^\alpha_{\mu\nu,\beta} + \Gamma^\sigma_{\mu\beta}\Gamma^\alpha_{\sigma\nu} - \Gamma^\sigma_{\mu\nu}\Gamma^\alpha_{\sigma\beta} \,. \tag{1.16}$$

The Riemannian tensor is skew-symmetric in the pair of indices

$$R_{\alpha\beta\mu\nu} = -R_{\alpha\beta\nu\mu} = -R_{\beta\alpha\mu\nu} \,. \tag{1.17}$$

The Riemannian tensor shows symmetry if we interchange two pairs of indices

$$R_{\alpha\beta\mu\nu} = R_{\mu\nu\alpha\beta} \,. \tag{1.18}$$

If the last three indices of the Riemannian curvature tensor are symmetrized, then due to symmetries of connection $\Gamma^\alpha_{\mu\nu}$, the sum vanishes:

$$R^\alpha_{[\beta\mu\nu]} = \frac{1}{3}\left(R^\alpha_{\beta\mu\nu} + R^\alpha_{\mu\nu\beta} + R^\alpha_{\beta\mu\nu}\right) = 0 \,. \tag{1.19}$$

The only nontrivial contraction of the Riemann curvature tensor (due to symmetries all others are zero) gives the Ricci tensor

$$R_{\mu\nu} = R^\alpha_{\mu\alpha\nu} \,, \tag{1.20}$$

which is a symmetric tensor. If we contract the Ricci tensor, the Ricci scalar is obtained:

$$R = g^{\mu\nu} R_{\mu\nu} \,. \tag{1.21}$$

1.1.3 Energy momentum tensor

The famous Newtonian formula for gravitational potential force is

$$\nabla^2 \Phi = 4\pi G \rho c^2 \,, \tag{1.22}$$

where Φ is the Newtonian gravitational potential, and ρ, c, and G are the energy density, the speed of light ($c = 2.9979 \times 10^8$ m/s in a vacuum) and gravitational constant

($G = 6.67 \times 10^{-11}\ \mathrm{Nm^2/kg^2}$) respectively. In general relativity theory, the quantity replacing ρ is the energy momentum tensor $T^{\mu\nu}$ and $\nabla^2\Phi$ is replaced by the Einstein tensor $G^{\mu\nu}$, which describes the curvature of spacetime.

Then, to get the general relativity version of equation (1.22), involving the energy momentum tensor, we just replace the Minkowski metric $\eta_{\alpha\beta}$ by the general metric tensor $g_{\alpha\beta}$, and the partial derivative (,) with the covariant derivative (;). In this framework, the energy momentum tensor of a perfect fluid in curved spacetime is

$$T^{\mu\nu} = \left(\rho + \frac{p}{c^2}\right) U^\mu U^\nu + p g^{\mu\nu}\,, \tag{1.23}$$

where p is the pressure and $U^\mu = (c, 0, 0, 0)$ is the four-velocity. Einstein realized that the way to construct the field equations was by finding a tensor that describes the curvature and whose divergence is zero. Tensor $T^{\mu\nu}$, with two indices, represents a symmetric second rank tensor, the same type as the Ricci curvature $R_{\mu\nu}$ tensor, which defines the curvature of spacetime. Then, the conservation law equations for the energy momentum tensor become

$$T^{\mu\nu}{}_{;\nu} = 0\,. \tag{1.24}$$

In general relativity it is possible to define all types of matter as perfect fluids; from stars to the whole universe. According to Weyl's Postulate, in general relativity the cosmological medium should be a perfect fluid. This indicates the existence of a comoving fluid with no particle intersections and no interactions. Therefore, we can take the perfect fluid tensor as our energy-momentum source.

1.1.4 Einstein field equations

Spacetime in general relativity is curved and its curvature is produced by active gravitational masses. The relation between curvature and mass is governed by Einstein's field equations. The vacuum Einstein field equations are

$$R^{\alpha\beta} = \Lambda g^{\alpha\beta}\,, \tag{1.25}$$

where Λ is the vacuum energy density, also called the cosmological constant. The vacuum shows lack of any matter, radiation, and other substances, and it has the lowest energy. As is seen in the field equation (1.25), even the lowest energy of vacuum causes spacetime to curve by affecting the gravitational field. If we choose the cosmological constant as zero then the vacuum equations are reduced to

$$R^{\alpha\beta} = 0\,. \tag{1.26}$$

Here the vacuum field equations (1.26) (or (1.25)) include ten nonlinear equations.

In the presence of matter, the Einstein equation changes to

$$G^{\mu\nu} = R^{\mu\nu} - \frac{1}{2} g^{\mu\nu} R + \Lambda g^{\mu\nu} = -\frac{8\pi G}{c^4} T^{\mu\nu}\,, \tag{1.27}$$

where $G^{\mu\nu}$ is called the Einstein tensor, which is a symmetric tensor.

1.2 The universe as a dynamical system

In general, the simplest structure of the universe is given by the metric tensor, whose components are solely time dependent. If we consider such an isotropic universe, which is characterized by time and the characteristic scale, then the evolution of that universe is described by the dynamical system. In a dynamical universe, the metric is defined by the Friedman–Robertson–Walker (FRW) metric and the dynamical system is described by the Friedman equations, which are constructed from the Einstein equations in the light of the FRW metric.

1.2.1 The Friedman–Robertson–Walker (FRW) metric and Friedman equations

Modern cosmological models are based on the cosmological principle, which states that the universe is isotropic and homogeneous everywhere. The metric of a homogeneous and isotropic universe, determined by this cosmological principle with cosmic time (t) and spatial coordinates (r, θ and ϕ), is the Robertson–Walker metric

$$ds^2 = -c^2 dt^2 + a^2(t)\left(dr^2 + R_c^2 S_k^2\left(\frac{r}{R_c}\right)\left[d\theta^2 + \sin^2\theta d\phi^2\right]\right) . \tag{1.28}$$

Here, r is the comoving radial distance, $R_c(t)$ is the radius of curvature at time t and by convention, R_c is the radius of the present universe. The function S_k is dependent on which geometry our universe has, specified through the index k. The index k is the normalized curvature constant and it is referred to as closed ($k > 0$), open ($k < 0$) or flat ($k = 0$). For S_k, depending on the geometry of the universe, the following expressions hold:

$$\begin{aligned} k = 1\,, \quad & S_{+1}\left(\frac{r}{R_c}\right) = \sin\left(\frac{r}{R_c}\right) , \\ k = 0\,, \quad & S_0\left(\frac{r}{R_c}\right) = \frac{r}{R_c} , \\ k = -1\,, \quad & S_{-1}\left(\frac{r}{R_c}\right) = \sinh\left(\frac{r}{R_c}\right) . \end{aligned}$$

Another useful quantity is the dimensionless expansion factor $a(t)$. It describes the expansion (or contraction) of the universe. It can be normalized in terms of its present epoch value $a(t_0) = 1$; at the moment of the Big Bang $a(t = 0) = 0$. The radius of curvature $R_c(t)$ also evolves along with the expansion factor $a(t)$. By definition, the normalized radius of curvature is given by

$$a(t) = \frac{R_c(t)}{R_c(0)} . \tag{1.29}$$

In 1922 Friedmann inferred the general Einstein equations for an isotropic and homogeneous medium. These Friedmann equations are given as

$$\left(\frac{\dot{a}}{a}\right)^2 = \frac{8\pi G}{3}\rho - \frac{kc^2}{R_0^2}\frac{1}{a^2} + \frac{\Lambda c^2}{3}, \tag{1.30}$$

$$\frac{\ddot{a}}{a} = -\frac{4\pi G}{3}\left(\rho + \frac{3p}{c^2}\right) + \frac{\Lambda c^2}{3}, \tag{1.31}$$

where $R_0 = R_{t_0}$ is the present-day value of the curvature radius, ρ and p are the energy density and the pressure of the universe, and Λ is the cosmological constant.

The expansion of the universe is given by the Hubble parameter

$$H = \frac{\dot{a}}{a}. \tag{1.32}$$

The present-day value of the Hubble parameter defines the Hubble constant. The recent value of the Hubble constant based on the analysis of the Hubble Space Telescope Key Project [20] is given by

$$H_0 = H(t_0) = 71 \pm 2\,, \quad \text{km s}^{-1}\ \text{Mpc}^{-1}. \tag{1.33}$$

For a given rate of expansion H, there is a critical density for which $k = 0$. The Wilkinson Microwave Anisotropy Probe (WMAP) results determined that the universe is flat, from which it follows that the mean energy density in the universe is equal to the critical density; that is, equivalent to a mass density

$$\rho_c = \frac{3H^2}{8\pi G} = 9.9 \times 10^{-30}\ \text{g/cm}^3. \tag{1.34}$$

The contribution of any energy component of the universe can be expressed in terms of the critical density by a dimensionless density parameter Ω

$$\Omega \equiv \frac{\rho(t)}{\rho_c}. \tag{1.35}$$

Due to contributions of the different components, the density parameter can be rearranged as

$$\Omega = \Omega_m + \Omega_r + \Omega_{DE}, \tag{1.36}$$

where the matter density has the contributions from the baryon and dark matter components $\Omega_m = \Omega_b + \Omega_{dm}$. Ω_r and Ω_{DE} show the contributions of radiation and dark energy components in the total density of the universe. Currently the favorite model is the concordance model or cold dark matter (CDM), with the following values:

$$\Omega_m = 0.279 \pm 0.025\,, \quad \Omega_r = 0.005\,, \quad \Omega_\Lambda = 0.721 \pm 0.025\,. \tag{1.37}$$

The values are based on the ninth annual data release of the WMAP [6]. The values of these fractional densities determine the global dynamics of the universe.

1.2.2 State equation and Friedman differential equation

In this section, we will discuss the universe as a thermodynamical system. We assume that the universe expands adiabatically, and that its adiabatic expansion satisfies the following equation:

$$dE + pdV = 0\,, \tag{1.38}$$

where V is the volume of a comoving sphere, which is proportional to the cube of the expansion factor $a(t)$, $V \sim a^3$. Here E is the total mass energy in the volume, equal to

$$E = \rho V = \rho a^3\,. \tag{1.39}$$

The first law of thermodynamics states that the change in internal energy ΔE of a system is equal to

$$\Delta E = \Delta Q - p\Delta V\,, \tag{1.40}$$

that is, the change in the heat ΔQ added to or escaped from the system, minus the work $\Delta W = p\Delta V$ done by the system. The particular case of adiabatic expansion occurs when no heat enters or leaves the system. This only happens when the work is done quickly. As a result, the adiabatic expansion indicates that there is no change in heat $\Delta Q = 0$. In this case, the equation of the first law of the thermodynamics (1.40) is reduced to the adiabatic expansion

$$\Delta E + p\Delta V = 0\,. \tag{1.41}$$

The above adiabatic expansion equation can be derived by using the dynamical equations of the universe; the so-called Friedman equations (1.30) and (1.31). As a result, one may verify that the universe expands adiabatically.

In a universe where adiabatic expansion occurs, there can be no thermal flow. In this case, the cosmic fluid obeys a simple equation of state

$$p := \omega c^2 \rho\,, \tag{1.42}$$

where the dimensionless number ω is called the adiabatic parameter. It is equal to:
(a) for pure radiation, $\omega = \frac{1}{3}$ (radiation/relativistic particle)
(b) for pure dust, $\omega = 0$ (dark matter/dust)
(c) for the cosmological constant, $\omega = -1$.

Simply put, the equation of state relates the pressure of the fluid to its energy density. In addition to this, equation (1.42) is particularly important because it allows us to obtain the exact solutions of the Friedman equations depending on the constituents of the universe.

2 Dynamics of universe models

The modern cosmology that we know was built by Nicolaus Copernicus in 1543. In his book *De revolutionibus orbium coelestium* (On the Revolutions of the Heavenly Spheres), he stated that there is no special observer. Based on this idea, Copernicus refused the Earth-centered model, and established the Sun-centered universe model instead.

His model described observed motions of celestial objects around the Sun. This heliocentric model became the most influential work in the history of science that led to modern cosmology. His model was developed and perfected by his scientific successors Tycho Brahe, Johannes Kepler, and Sir Isaac Newton, who helped explain the force of gravity that all bodies exert on each other. Copernicus's no special observer statement, in which the Sun was of course the preferred observer, became a principle called the cosmological principle.

The application of the cosmological principle to the universe indicates two important mathematical properties: that the universe should be homogeneous and isotropic. Isotropy applies at some specific points in space, where space looks the same independent of the direction you look in. It is isotropy that is indicated by the observations of the cosmic microwave background. The homogeneity property states that the metric is the same everywhere in space.

Based on the Copernicus's cosmological principle, the construction of cosmological models begins with the idea that the universe is homogeneous and isotropic in space on large scales. In general relativity this means that the universe can be presented by space-like slices that are homogeneous and isotropic.

2.1 The Friedman models

In this section we will discuss the solutions of the isotropic and homogeneous dust/dark matter universe models derived from the Friedman equation (1.30). These dust solutions of the Friedman differential equation are also known as Friedman universe models. As is known, in the dust dominated universe, the density function is proportional to expansion factor $\rho \propto a^{-3}$. So, if we rearrange equation (1.30), then we obtain the following form:

$$\dot{a}^2 = \frac{8\pi G}{3}\rho a^2 + \frac{\Lambda c^2}{3}a^2 - kc^2 =: F(a)\,, \tag{2.1}$$

where $F(a)$ is the function of the scale factor. This form of the Friedman differential equation (2.1) allows us to obtain the time parameter easily by applying the separation of variables method

$$t = \int \frac{da}{\sqrt{F(a)}}\,, \quad a > 0\,. \tag{2.2}$$

https://doi.org/10.1515/9783110515367-002

The Friedman dust models are obtained by using the Friedman equation (2.1) along with the time equation (2.2). Then the dust models can be classified as
- Static models
- Empty models
- Non-empty models with $\Lambda = 0$
- Non-empty models with $\Lambda \neq 0$ [35].

In the following subsections, we derive the mathematical formulations of the above universe models, which evolve from very simple to complex forms due to the addition of nonzero gravity and a nonzero cosmological constant.

2.1.1 The static models

Static models do not show dynamical evolution satisfying constant expansion a and constant density function ρ. Therefore, in static models, the Friedman equation (2.1) is reduced to the following form:

$$\frac{\kappa}{a^2} = \frac{8\pi G\rho}{3c^2} + \frac{\Lambda}{3} \,. \tag{2.3}$$

There are two universe models that can be constructed from the above equation. These are:
- Einstein universe ($\rho > 0$ and $\kappa = 1$):
  This model was proposed by Einstein in 1917. As is seen from equation (2.3), this model cannot stay in static equilibrium due to positive constant energy density ρ, which creates a source of gravity. That is why the cosmological constant Λ on the right-hand side of the equation (2.3) has crucial importance.
- Static nongravitating universe ($\rho = \Lambda = \kappa = 0$ and a =constant):
  Under these conditions, the model represents a static, nongravitating universe due to the lack of a source of gravity $\rho = 0$. As a result, this model can be represented by the Minkowski metric.

2.1.2 Empty models

Assuming zero density $\rho = 0$, the gravity of the universe model is switched off. This leads to the Friedman equation (2.1), reduced to the following form:

$$\dot{a}^2 = \frac{\Lambda c^2}{3} a^2 - \kappa c^2 = F(a) \,. \tag{2.4}$$

Using the above equation, we can derive different empty models with different time parameters determined by equation (2.2), which appears in a simple form, allowing us to integrate it analytically

$$t = \int \frac{da}{\sqrt{F(a)}} = \int \frac{1}{\sqrt{\frac{\Lambda a^2}{3} - \kappa}} da\,. \tag{2.5}$$

Hence, the empty models are the following;

(a) If $\Lambda = 0$, $\kappa = -1$, then the time equation (2.5) is reduced to

$$t = \int \frac{da}{c}\,, \tag{2.6}$$

which results in $a = ct$, also known as Milne's model. Note that the conditions $\Lambda = 0$ and $\kappa = -1$ provide the only analytical solution for real $t \in \mathbb{R}$. Supposing $\Lambda = 0$ with positive or zero curvature will lead to complex or divergent time parameters respectively.

(b) If $\Lambda > 0$ and $\kappa = 0$, then the time equation (2.5) is reduced to

$$t = \sqrt{\frac{3}{\Lambda}} \frac{1}{c} \int \frac{da}{a}\,, \tag{2.7}$$

which leads to the expansion factor

$$a = e^{\sqrt{\frac{\Lambda}{3}} ct}\,, \tag{2.8}$$

determining the expansion rate in the de Sitter space.

(c) If $\Lambda > 0$, $\kappa = 1$, then equation (2.5) becomes

$$t = \int \frac{da}{c\sqrt{\frac{\Lambda}{3} a^2 - 1}}\,. \tag{2.9}$$

The solution of this integral comes from using the hyperbolic substitution, $a = \sqrt{\frac{3}{\Lambda}} \cosh\theta$. Then the expansion factor is

$$a(t) = \sqrt{\frac{3}{\Lambda}} \cosh\left(\sqrt{\frac{\Lambda}{3}} ct\right)\,. \tag{2.10}$$

This solution for the expansion factor, satisfying conditions $\Lambda > 0$ and $\kappa = 1$, also gives the de Sitter-type expansion.

(d) If $\Lambda > 0$ and $\kappa = -1$, then the integral in equation (2.5) is reduced to the following:

$$t = \int \frac{da}{c\sqrt{\frac{\Lambda}{3} a^2 + 1}}\,. \tag{2.11}$$

By using the hyperbolic substitution, $a = \sqrt{\frac{3}{\Lambda}} \sinh\theta$, one can obtain the expansion factor in the hyperbolic form as in the previous model

$$a(t) = \sqrt{\frac{3}{\Lambda}} \sinh\left(\sqrt{\frac{\Lambda}{3}} ct\right)\,. \tag{2.12}$$

This hyperbolic solution of the scale factor $a(t)$ also gives de Sitter-type expansion.

(e) If $\Lambda < 0$ and $\kappa = -1$, then the integral (2.5) becomes

$$t = \int \frac{da}{c\sqrt{1 - \frac{\Lambda}{3}a^2}} . \tag{2.13}$$

Applying trigonometric substitution, $a = \sqrt{\frac{3}{\Lambda}} \sin\theta$, one can obtain the solution as

$$a(t) = \sqrt{\frac{3}{\Lambda}} \sin\left(\sqrt{\frac{\Lambda}{3}} ct\right) . \tag{2.14}$$

This model provides an analogy of Milne's model in anti-de Sitter space.

Figure 2.1 shows the evolution of the expansion factors of the empty models in terms of time. As is seen, the presence of the positive cosmological constant $\Lambda > 0$ leads to the empty models with exponentially expanding scale factors. In these models ($\Lambda > 0$), the exponential growth of the scale factors is not strongly affected by the geometry. However, a negative cosmological constant and curvature are forcing the scale factor to show periodic behavior. In the case of a zero cosmological constant, the scale factor is proportional to time with the velocity of light.

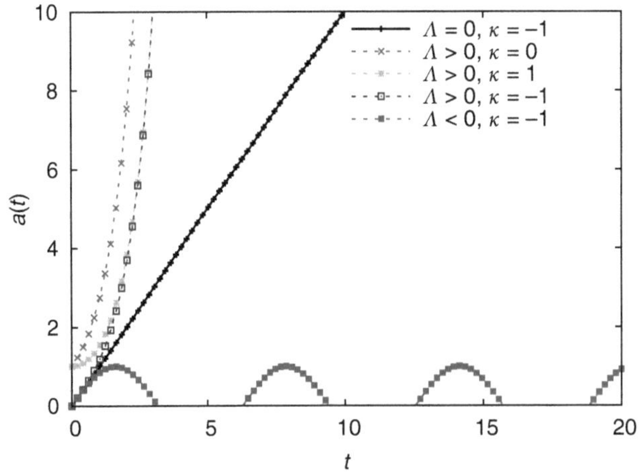

Fig. 2.1: The scale factors $a(t)$ of the different empty models versus time, t. Note that the velocity of light and cosmological constant are assumed to be unity.

2.1.3 Non-empty models with $\Lambda = 0$

The non-empty models assume a nonzero density function and a zero cosmological constant. Under these conditions, the Friedman equation is reduced to

$$F(a) = \frac{8\pi G}{3a} - \kappa c^2 , \tag{2.15}$$

assuming again the dust dominated case, where $\rho \propto a^{-3}$. This differential equation allows us to construct three non-empty models due to the curvature parameter κ. The curvature admits three possible geometries: -1, 0, and 1.

(a) $\Lambda = 0$ and $\kappa = 0$

In the case of the flat geometry with a zero cosmological constant, the dust Friedman equation (2.15) is simplified to the following form:

$$F(a) = \frac{8\pi G}{3a} . \tag{2.16}$$

Following this, equation (2.2) becomes

$$t = \sqrt{\frac{3}{8\pi G}} \int \sqrt{a}\, da = \sqrt{\frac{1}{6\pi G}} a^{3/2} .$$

This non-empty model without a cosmological constant contribution is known as the Einstein–de Sitter model, in which the scale factor a evolves as

$$a \propto t^{2/3} .$$

Moreover, one can obtain the density parameter of the Einstein–de Sitter model by using equation (2.16) as follows:

$$\rho = \frac{3}{8\pi G} H^2 , \tag{2.17}$$

where $H = \frac{d}{dt} \ln a$ is called the Hubble parameter.

(b) $\Lambda = 0$ and $\kappa = 1$

In the case of closed geometry – the non-empty universe with $\Lambda = 0$ – the Friedman equation becomes

$$F(a) = c^2 \left(\frac{K}{a} - 1 \right) , \tag{2.18}$$

in which $K \equiv \frac{8\pi G}{3c}$. Hence the time parameter given by equation (2.2) takes the following form:

$$t = \int \frac{\sqrt{a}}{c\sqrt{K - a}} da .$$

In the above integral, by using the trigonometric substitution $a = K \sin^2 \theta$, $da = 2K \sin\theta \cos\theta$, we obtain the following solution of the time parameter in terms of the scale factor:

$$t = \frac{K}{c} \left[\arcsin \sqrt{\frac{a}{K}} - \sqrt{\frac{a}{K} - \frac{a^2}{K^2}} \right] . \tag{2.19}$$

As is seen from above form of the time parameter, it is difficult to obtain the scale factor in terms of the time parameter. To do this, one can change variables and rearrange the time equation (2.19) in parametric form. For example, assuming $\frac{a}{K} = \sin^2 \frac{\chi}{2}$ in Equation (2.19), one can obtain the following form:

$$t = \frac{K}{c}\left[\frac{\chi}{2} - \sin\frac{\chi}{2}\sin\frac{\chi}{2}\right] . \tag{2.20}$$

Using the half-angle formula $\sin\chi = 2\sin\frac{\chi}{2}\sin\frac{\chi}{2}$ and taking the derivative of both sides in terms of χ in the above equation, one can obtain the parametric equation of a cycloid

$$\frac{dt}{d\chi} = \frac{K}{2c}[1 - \cos\chi] = a , \quad \chi \in [0, 2\pi] ,$$

for the closed model with $\kappa = 1$. The maximum radius of the cycloid is given by $\chi = \pi$ while the total duration becomes $t_{tot} = t(\chi = 2\Pi) = \pi C$.

(c) $\Lambda = 0$ and $\kappa = -1$

When the geometry is defined by the negative curvature, the non-empty models with zero cosmological constant are given by

$$F(a) = c^2\left(\frac{K}{a} + 1\right) , \tag{2.21}$$

which leads to the time parameter

$$t = \int \frac{\sqrt{a}}{\sqrt{K + a}} da . \tag{2.22}$$

To obtain the solution of the above integral, the hyperbolic substitution $a = K\sinh^2\theta$ can be used. Then the solution becomes

$$t = K\left[\sqrt{\frac{a}{K} + \frac{a^2}{C^2}} - \operatorname{arcsinh}\sqrt{\frac{a}{C}}\right] . \tag{2.23}$$

By following a similar procedure as in the closed geometry, the parametric form of the negatively curved non-empty models can be obtained. First it is assumed that $X = \frac{a}{K}$, and then change of variables $X = \sinh^2\frac{\chi}{2}$ is used in the time equation (2.23):

$$\frac{dt}{d\chi} = \frac{K}{2c}[\cosh\chi - 1] = a . \tag{2.24}$$

The integration of this equation in terms of χ gives the time parameter as

$$t = \frac{K}{2c}[\sinh\chi - \chi] . \tag{2.25}$$

Figure 2.2 presents the two scale factors of the non-empty closed and open models. As is seen, the existence of the density in the models as a gravitational source causes different behaviors of the parametrized scale factors depending on the geometry. While the open non-empty model shows an exponentially growing scale factor, the scale factor of the non-empty closed model becomes periodic by satisfying the parametrized cycloid equation.

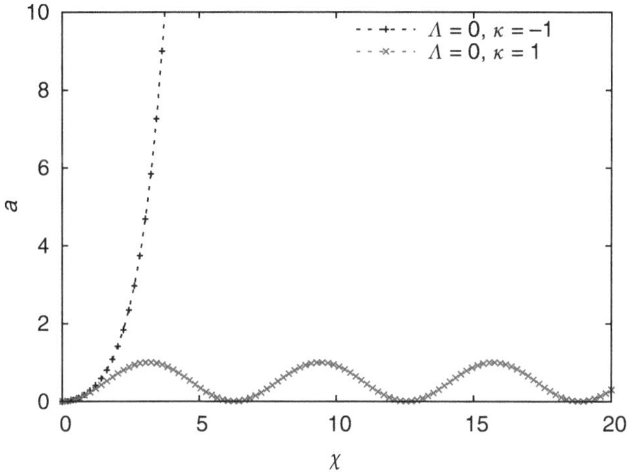

Fig. 2.2: The scale factors $a(\chi)$ of the two non-empty models for closed and open geometries.

2.1.4 Non-empty models with $\Lambda \neq 0$

If we take into account the arbitrary value of Λ, then the Friedman differential equation for the dust models becomes

$$\dot{a}^2 + \kappa c^2 = c^2\left(\frac{K}{a} + \frac{\Lambda a^2}{3}\right) =: F(a, \Lambda)\,. \tag{2.26}$$

Here $F(a, \Lambda)$ is a multivariable function.

2.2 Milne's model

Milne's model is an empty universe corresponding to the limit $\rho \longrightarrow 0$ with $\Lambda = 0$. Using Friedman equation (2.26) for three different curvatures – $\kappa = 0, 1$, and -1 – Milne's model becomes the reduction of equation (2.26):

$$\dot{a}^2 = -\kappa c^2\,. \tag{2.27}$$

- In flat geometry ($\kappa = 0$), equation (2.27) gives a constant scale factor, $a(t) = constant$.
- In closed geometry ($\kappa = 1$), equation (2.27) gives a complex solution, $a(t) = ict$.
- In hyperbolic geometry ($\kappa = -1$), equation (2.27) gives a linear relation between the scale factor and time parameter, $a(t) = ct$.

Here it is important to mention that the FRW metric reduces to the Milne universe in the case of a vacuum without matter content and cosmological constant. In this case,

Milne provides a cosmological solution to the Einstein field equations with the metric

$$ds^2 = c^2 dt^2 - dr^2 - r^2 \left(d\theta^2 + \sin^2\theta d\phi^2\right) , \tag{2.28}$$

in which the scale factor is constant over time, and r gives the proper motion distance between points in the universe. Therefore, the flat FRW metric becomes the Milne metric, which is a restatement of the Minkowski metric.

3 Anisotropic and homogeneous universe models

Providing generalizations of the Friedman–Robertson–Walker models, the Bianchi universe models are spatially homogeneous and anisotropic. These anisotropic models are classified into nine different types and two main classes by Luigi Bianchi [7] as simply transitive groups of isometries G_3 by using three-dimensional Lie algebras. These isometries act on space-like three-surfaces of the metric tensor and make these surfaces homogeneous, since cosmological time is constant (t = const). The isometry group G_3 provides a simple representation of the universe models based on the existence of sets of four mutually orthogonal basis vector fields. These vector fields are known as tetrads e_i ($i = 0, 1, 2, 3$) and are invariant under all transformations of the isometry group G_3.

Apart from the tetrad representation of the Bianchi classification, it is useful to give the well-known metric representations of the Bianchi types:

- Type I: $ds^2 = c^2dt^2 - a_1^2dx^2 - a_2^2dy^2 - a_3^2dz^2$
- Type II: $ds^2 = c^2dt^2 - a_1^2\left(dx^2 + dy^2\right) - a_2^2\left(dz + xdy\right)^2$
- Type III: $ds^2 = c^2dt^2 - a_1^2dx^2 - a_2^2e^{-2x}dy^2 - a_3^2dz^2$
- Type IV: $ds^2 = c^2dt^2 - a_1^2dx^2 - a_2^2dy^2 - a_3^2e^{2y}\left(ydx - dz\right)^2$
- Type V: $ds^2 = c^2dt^2 - a_1^2dx^2 - a_2^2e^{2x}\left(dy^2 + dz^2\right)$
- Type VI_h: $ds^2 = c^2dt^2 - a_1^2e^{2(h-1)y}dx^2 - a_2^2dy^2 - a_3^2e^{2(h+1)y}dz^2$
- Type VII_h: $ds^2 = c^2dt^2 - a_1^2e^{2hy}dx^2 - a_2^2dy^2 - a_3^2e^{2(h+1)y}dz^2$
- Type VIII: $ds^2 = c^2dt^2 - a_1^2\left(dx^2 + \cosh^2 xdy^2\right) - a_2^2\left(dz + \sinh xdy\right)^2$
- Type IX: $ds^2 = c^2dt^2 - a_1^2\left(dx^2 + \sin^2 xdy^2\right) - a_2^2\left(dz + \cos xdy\right)^2$,

where a_1, a_2 and a_3 are the expansion factors of the three-dimensional space of the Bianchi models. Some of these models admit FRW-type solutions, such as I and VII_0 for a flat geometry ($k = 0$), V and VII_h for an open geometry ($k = -1$), and IX for a closed geometry ($k = +1$). The Bianchi models that do not admit FRW solutions tend to be highly anisotropic. Here only the Bianchi Type I (BI) models will be discussed in detail.

3.1 Bianchi type I models

The homogeneity and isotropy of the FRW models make them the simplest possible models for a uniformly expanding universe. As is indicated by [16], the main reason for studying these models has been their mathematical simplicity rather than any observational evidence. However, the cosmic microwave background (CMB) observations show that the universe on large scales is very close to that of a FRW model. It is still possible to choose homogeneous models that are initially anisotropic but become more isotropic as time goes on, and asymptotically tend to a FRW model. As a result, here we

https://doi.org/10.1515/9783110515367-003

will consider models of a universe with an initially anisotropic background. The more general anisotropic cases are Bianchi Type I homogeneous models whose spatial sections are flat but whose expansion rates are direction dependent [38]. The geometry of the Bianchi Type I model admits the metric element

$$dl^2 = c^2 dt^2 - a_1^2(t)dx^2 - a_2^2(t)dy^2 - a_3^2(t)dz^2 \tag{3.1}$$

that has three different scale factors – a_1, a_2 and a_3 – that are a function of time t. Taking into account the energy-momentum tensor of a perfect fluid, the Einstein field equations of the BI universe become

$$\frac{\dot a_1 \dot a_2}{a_1 a_2} + \frac{\dot a_1 \dot a_3}{a_1 a_3} + \frac{\dot a_2 \dot a_3}{a_2 a_3} = 8\pi G\rho\,, \tag{3.2a}$$
$$\frac{\ddot a_1}{a_1} + \frac{\ddot a_3}{a_3} + \frac{\dot a_1 \dot a_3}{a_1 a_3} = -\frac{8\pi G}{c^2}p\,, \tag{3.2b}$$
$$\frac{\ddot a_2}{a_2} + \frac{\ddot a_1}{a_1} + \frac{\dot a_2 \dot a_1}{a_2 a_1} = -\frac{8\pi G}{c^2}p\,, \tag{3.2c}$$
$$\frac{\ddot a_3}{a_3} + \frac{\ddot a_2}{a_2} + \frac{\dot a_3 \dot a_2}{a_3 a_2} = -\frac{8\pi G}{c^2}p\,, \tag{3.2d}$$

in which the dot represents the derivatives in terms of time t. To solve the System (3.2), we define the following new variables, which are simply the directional Hubble parameters:

$$H_1 \equiv \frac{\dot a_1}{a_1}\,, \quad H_2 \equiv \frac{\dot a_2}{a_3}\,, \quad H_3 \equiv \frac{\dot a_3}{a_3}\,. \tag{3.3}$$

Inserting variables (3.3) and their first derivatives into the Einstein field equations (3.2), we reformulate the field equations in terms of the directional Hubble parameters:

$$H_1 H_2 + H_1 H_3 + H_2 H_3 = 8\pi G\rho\,, \tag{3.4a}$$
$$\dot H_3 + H_3^2 + \dot H_1 + H_1^2 + H_3 H_1 = -\frac{8\pi G}{c^2}p\,, \tag{3.4b}$$
$$\dot H_1 + H_1^2 + \dot H_2 + H_2^2 + H_1 H_2 = -\frac{8\pi G}{c^2}p\,, \tag{3.4c}$$
$$\dot H_3 + H_3^2 + \dot H_2 + H_2^2 + H_3 H_2 = -\frac{8\pi G}{c^2}p\,. \tag{3.4d}$$

In addition to this, the energy-momentum conservation equation $T^{\mu}_{\nu;\mu} = 0$ yields

$$\dot\rho = -\left(\frac{\dot a_1}{a_1} + \frac{\dot a_2}{a_2} + \frac{\dot a_3}{a_3}\right)\left(\rho + \frac{p}{c^2}\right) = -3H\left(\rho + \frac{p}{c^2}\right)\,. \tag{3.5}$$

As is known, the BI universe has a flat metric with $k = 0$, which implies that its total density is equal to the critical density. The critical density is given by

$$\rho = \rho_c = \frac{1}{8\pi G}\left(H_1 H_2 + H_1 H_3 + H_2 H_3\right)\,. \tag{3.6}$$

3.1.1 General solution

In this subsection, we derive the analytical solutions of the field equations of the BI models in terms of the directional Hubble parameters. To do this, first we add the last three equations of system (3.4), which yields

$$2\frac{d}{dt}\left(\sum_{i=1}^{3} H_i\right) + 2\left(H_1^2 + H_2^2 + H_3^2\right) + (H_3H_2 + H_1H_2 + H_3H_1) = \frac{-24\pi G}{c^2}p\,. \tag{3.7}$$

After substituting the following term:

$$\sum_{i=1}^{3} H_i^2 = \left(\sum_{i=1}^{3} H_i\right)^2 - 2\,(H_3H_2 + H_1H_2 + H_3H_1)\;, \tag{3.8}$$

and equation (3.4a) of System (3.4) into equation (3.7), we then obtain

$$\frac{d}{dt}\left(\sum_{i=1}^{3} H_i\right) + \left(\sum_{i=1}^{3} H_i\right)^2 = 12\pi G\left(\rho - \frac{p}{c^2}\right)\,. \tag{3.9}$$

The mean of the three directional Hubble parameters in the BI universe is given by

$$H \equiv \frac{1}{3}\,(H_1 + H_2 + H_3) = \frac{1}{3}\left(\frac{\dot a_1}{a_1} + \frac{\dot a_2}{a_2} + \frac{\dot a_3}{a_3}\right)\,. \tag{3.10}$$

Substituting the mean (3.10) into equation (3.9), a nonlinear first-order differential equation is obtained

$$\dot H + 3H^2 = 4\pi G\left(\rho - \frac{p}{c^2}\right)\,. \tag{3.11}$$

Here, this dynamical equation shows evolution of the Hubble parameter of the related BI cosmology. Mathematically, this equation is the first-order nonlinear differential equation known as the Riccati equation. It can be linearized in terms of the second-order linear equation by special substitution.

Indeed, it is possible to write equation (3.11) in terms of volume element V by using the following relation between volume and the mean Hubble parameter of the BI:

$$H = \frac{1}{3}\frac{d}{dt}\ln(a_1a_2a_3) = \frac{1}{3}\frac{\dot V}{V}\,. \tag{3.12}$$

As is seen, the multiplication of the scale factors in different directions is defined as the volume element of the BI universe $V \equiv a_1a_2a_3$. Using this relation between volume and the mean Hubble parameter in equation (3.11), the volume evolution equation of the BI models is obtained

$$\ddot V - 3\left[4\pi G\left(\rho - \frac{p}{c^2}\right)\right]V = 0\,. \tag{3.13}$$

On the basis of the above, we find the following alternative form for System (3.4):

$$\dot{H}_1 + 3H_1H = 4\pi G\left(\rho - \frac{p}{c^2}\right), \tag{3.14a}$$

$$\dot{H}_2 + 3H_2H = 4\pi G\left(\rho - \frac{p}{c^2}\right), \tag{3.14b}$$

$$\dot{H}_3 + 3H_3H = 4\pi G\left(\rho - \frac{p}{c^2}\right). \tag{3.14c}$$

These expressions allow us to write down the generic solution of the directional Hubble parameters

$$H_i(t) = \frac{1}{\mu(t)}\left[K_i + \int \mu(t)4\pi G\left(\rho(t) - \frac{p(t)}{c^2}\right)dt\right], \quad i = 1, 2, 3, \tag{3.15}$$

where K_is are the integration constants. The integration factor μ is defined as

$$\mu(t) = e^{\int^t 3H(s)ds}. \tag{3.16}$$

The integration factor μ in the solutions (3.15) is derived from the system (3.14) by the particular solution of the system itself.

As can be seen from the solutions (3.14), the initial values/integration constants determine the solution of each directional Hubble parameter. These values are the origin of the anisotropy. Note that the generic solution of the directional Hubble parameters (3.15) is incomplete. To obtain exact solutions of the Hubble parameters and therefore the Einstein equations, we need one more equation, which is known as the equation of state (1.42).

As was stated before, the isotropic and homogeneous nature of the large-scale structure of the universe may be an asymptotic situation emerging from an anisotropic expansion in the very early universe. That is why it is important to define an isotropization criteria. This should explain how the anisotropy parameters disappear or become negligible when the universe evolves into the present epoch, $t \to t_0$. [10, 39, 40] define isotropization as expansion factors of the BI universe that grow at the same rate at later stages of the evolution. It is assumed that a BI model becomes isotropic if the ratio of each directional expansion factor $a_i(t)$ $(i = 1, 2, 3)$ and the expansion factor of the total volume $a(t)$ tends to be a constant value C_i,

$$\frac{a_i}{a} \to \text{constant} > 0 \text{ when } t \to \infty. \tag{3.17}$$

Note that the total expansion factor $a(t)$ has the contribution from each directional expansion factor

$$a = (a_1a_2a_3)^{1/3} = V^{1/3}. \tag{3.18}$$

The anisotropic models satisfying condition (3.17) become isotropic. As a particular case of condition (3.17), one may choose the constant as unity when time tends to be identical and equivalent to unity in the limit of present-day $t = t_0$. The reason for choosing this constant as unity at the present day is to construct Bianchi models that satisfy isotropy as the FRW ones do, even though they are highly anisotropic at the beginning of their evolution. Although this statement is correct, it is insufficient to construct a realistic model that starts its evolution highly anisotropic then evolves into the FRW universe. It should be specifically indicated that the anisotropic parts of each scale factor of the BI models should tend to be identical and equivalent to unity in the limit of present-day $t = t_0$. Under this condition, the critical density of the BI models is reduced to

$$\frac{\rho}{\rho_c} = \Omega = 1\,, \quad \rho_c = \frac{3H_0^2}{8\pi G}\,, \tag{3.19}$$

where the total mean density Ω of the Bianchi models is unity due the flat geometry of the Bianchi metric (3.1). Apart from this general isotropization criterion (3.17), we consider the following isotropization that comes from the consistency relation of the analytical solution of the field equations of the BI metric (3.2) via the integration constants [10, 25, 34, 39, 40]:

$$\sum_{i=1}^{3} K_i = 0\,. \tag{3.20}$$

3.1.2 Sample solution: radiation dominated Bianchi type I model

Here we assume the universe emerges out of anisotropy in the radiation dominated epoch and investigate how this anisotropy turns into isotropy depending on the anisotropy coefficients. In the radiation dominated BI universe the pressure term is proportional to density ρ_r with $\gamma = 1/3$, similar to the radiation dominated FRW model. Since the radiation is the dominant component, we take into account the presence of only radiation in the equation of state (1.42). By using the volume representation of the mean Hubble parameter from relation (3.12), the energy density of the radiation dominated BI universe in terms of volume element V_r is obtained as

$$\rho_r = \rho_{r,0}\left(\frac{V_{r,0}}{V_r}\right)^{4/3}\,, \tag{3.21}$$

in which the density and the volume element is normalized to the present-day t_0. Then parameters $\rho_{r,0}$ and $V_{r,0}$ are the normalized density and normalized volume elements respectively. To obtain the exact solution of system (3.15), first the form of the mean Hubble parameter H_r is obtained by substituting the equation of state (1.42) of the radiation dominated epoch in (3.11), which leads to

$$\dot{H}_r + 3H_r^2 = \frac{8\pi G}{3}\rho_r\,. \tag{3.22}$$

The overall volume evolution of the model is then obtained by substituting the density ρ_r from equation (3.21) and relation (3.12) in the mean Hubble parameter evolution equation (3.22)

$$\ddot{V}_r - 8\pi G\rho_{r,0}V_r^{-1/3} = 0\,. \tag{3.23}$$

Multiplying equation (3.23) with the time derivative of the volume element $\dot{V}_r$ and integrating it yields

$$\dot{V}_r^2 - 24\pi G\rho_{r,0}V_r^{2/3} = 0\,, \tag{3.24}$$

in which $\rho_{r,0} = \rho_{c,0}\Omega_{r,0} = 9H_0^2\Omega_{r,0}$ from the description of the critical density (3.6). Equation (3.24) is equal to curvature of the model and its value is zero since the BI universe has a flat geometry $\Omega = \Omega_{r,0} = 1$. Hence, the exact solution of the volume evolution equation can be obtained easily:

$$V_r = (2H_0t)^{3/2}\,. \tag{3.25}$$

Substituting this volume element (3.25) into equation (3.12), the mean Hubble parameter of the radiation dominated case is found as

$$H_r = \frac{1}{2}\frac{1}{t}\,. \tag{3.26}$$

Hence the integration factor to obtain the directional expansion rates is obtained from definition (3.16) by direct substitution of the mean Hubble parameter (3.26) of the radiation dominated model, which is

$$\mu_r = t^{3/2}\,. \tag{3.27}$$

By direct substitution of the integration factor (3.27) and the equation of state (1.42) of the radiation dominated case into the solutions (3.15), the directional Hubble parameters that are normalized to the present-day t_0 are given by

$$H_{r,i}t_0 = \alpha_{r,i}\left(\frac{t_0}{t}\right)^{3/2} + \frac{1}{2}\frac{t_0}{t}\,, \tag{3.28}$$

where $\alpha_{r,i}$s are the normalized anisotropy coefficients and are defined as

$$\alpha_{r,i} \equiv \frac{K_{r,i}}{\sqrt{t_0}}\,. \tag{3.29}$$

The normalized scale factors are derived from the directional Hubble parameters (3.15) with a direct integration in terms of cosmic time, which are obtained as

$$a_{nr,i} = \underbrace{e^{-2\alpha_{r,i}\left(\sqrt{\frac{t_0}{t}}-1\right)}}_{a_{n,i}}\underbrace{\left(\frac{t}{t_0}\right)^{1/2}}_{a_{nrFRW}}\,. \tag{3.30}$$

Here index n stands for the normalization of the scale parameters to the present-day t_0. The scale factors of the BI radiation dominated model has the contribution from anisotropic expansion/contraction $a_{n,i}$ as well as the standard matter dominated FRW model $a_{nr,FRW}$. These two different dynamical behaviors in three directional scale factors of the BI universe indicate that the FRW part of the scale factor becomes dominant when time approaches the present day. In contrast, in the early times of the BI model the expansion is completely dominated by the anisotropic part $a_{n,i}$. During the anisotropic expansion period of the BI model, the directional scale factors can show contraction and/or expansion behavior until a specific time when isotropy overcomes anisotropy and the directional scale factors become equivalent. In this study, we provide critical anisotropy coefficients $\alpha_{r,i}$ in the BI radiation dominated model, which are

$$\alpha_{r,i} = -\frac{1}{2}\sqrt{\frac{t}{t_0}} \,. \tag{3.31}$$

These critical values are obtained from the first derivative of the scale factors (3.30) by taking this derivative equivalent to zero in order to find possible turning behaviors of the scale factors. This mathematical description provides possible contracting or expanding dynamical characteristics of the scale factors during the initial anisotropic evolution of the BI model. When we investigate the behavior of these critical anisotropy coefficients (3.31), we find the criteria on the anisotropy coefficients, which provide the information at which values the scale factors show expansion and contraction initially in the radiation dominated BI model. These criteria are given by

$$\text{Expansion } \alpha_{r,i} = \begin{cases} \text{If } \alpha_{r,i} < 0, \text{ then } |\alpha_{r,i}| < \frac{1}{2}\sqrt{\frac{t}{t_0}} \,, \\ \text{If } \alpha_{r,i} > 0, \text{ then } \alpha_{r,i} > -\frac{1}{2}\sqrt{\frac{t}{t_0}} \,. \end{cases} \tag{3.32a}$$

$$\text{Contraction } \alpha_{r,i} = \left\{ \text{If } \alpha_{r,i} < 0 \text{ then } |\alpha_{r,i}| > \frac{1}{2}\sqrt{\frac{t}{t_0}} \,. \right. \tag{3.32b}$$

These initial expansions and/or contractions of the scale factors affect the dynamical behaviors of the directional Hubble parameters as expected. A possible contraction of one of the directional scale factors causes slowing down in the rate of expansion, which leads to zero or even negative expansion rates depending on how strong the slowing down is in the related directions. Criteria (3.32) therefore indicate additional criteria for the directional Hubble parameters by presenting emergence out of positive (same as expansion criterion (3.32a)) and negative (same as contraction criterion (3.32b)) branches.

In Figures 3.1 and 3.2, the change in dynamical behaviors of the directional scale factors and the directional Hubble parameters are presented using the different sets of anisotropy coefficients chosen from the criteria (3.32) that satisfy the consistency relation (3.20). In Figures 3.1 and 3.2, the first panels with zero anisotropy coefficients $\alpha_{r,1} = \alpha_{r,1} = \alpha_{r,3} = 0$ (top left in each figure) present the evolution of the standard FRW parameters for comparison with BI models. Additionally, in Figures 3.1 and 3.2, the anisotropic coefficients are specially chosen to show the evolution of the dynamical parameters around and at the critical value $-1.387 \cdot 10^{-4}$ from equation (3.31) by choosing the initial time as $t/t_0 \approx 7.7 \cdot 10^{-8}$ ($\approx 10^{-34}$ seconds) when the radiation dominated era starts. Figure 3.2 shows the emergence out of a negative branch in

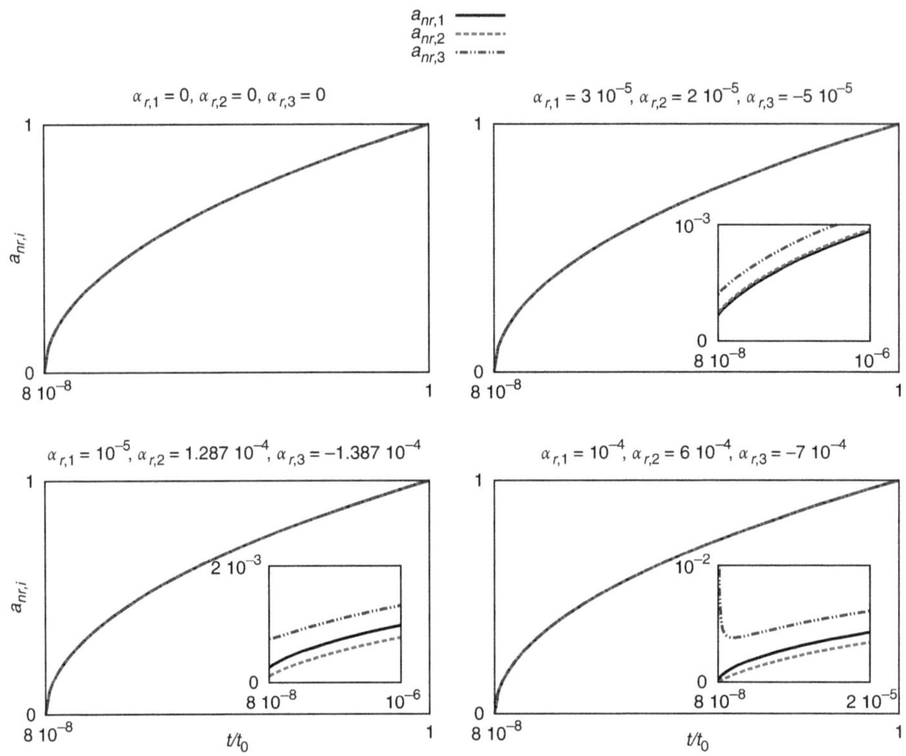

Fig. 3.1: Evolution of the normalized scale factors starting at $t/t_0 = 7.7 \cdot 10^{-8}$ in terms of different sets of anisotropy coefficients.

the third directional Hubble parameter $H_{r,3}$. It is evident that when the model dependent anisotropy coefficient $\alpha_{r,3}$ reaches the critical value $-1.387 \cdot 10^{-4}$, the related Hubble parameter accepts value zero (bottom left in Figure 3.2). This means that the related scale continues its evolution constantly without accelerating or decelerating (bottom left in Figure 3.1). When the $\alpha_{r,3}$ accepts the criterion (3.32b) to obtain contraction $|\alpha_{r,3}| > 1.387 \cdot 10^{-4}$, depending on the absolute value of $|\alpha_{r,3}|$, one may obtain initially small or very large negativity in the directional Hubble parameter (bottom right panel in Figure 3.2), while this indicates that the related scale factor contracts (bottom right panel in Figure 3.1).

The anisotropy coefficients $\alpha_{r,i}$ have key importance in modeling an initially anisotropic and asymptotically FRW model. It is crucial to note that even though the BI model shows initial contraction or expansion in each direction, the overall volume of the universe behaves as in the standard FRW model. This can be proven by a simple mathematical calculation; the multiplication of the three normalized scale factors,

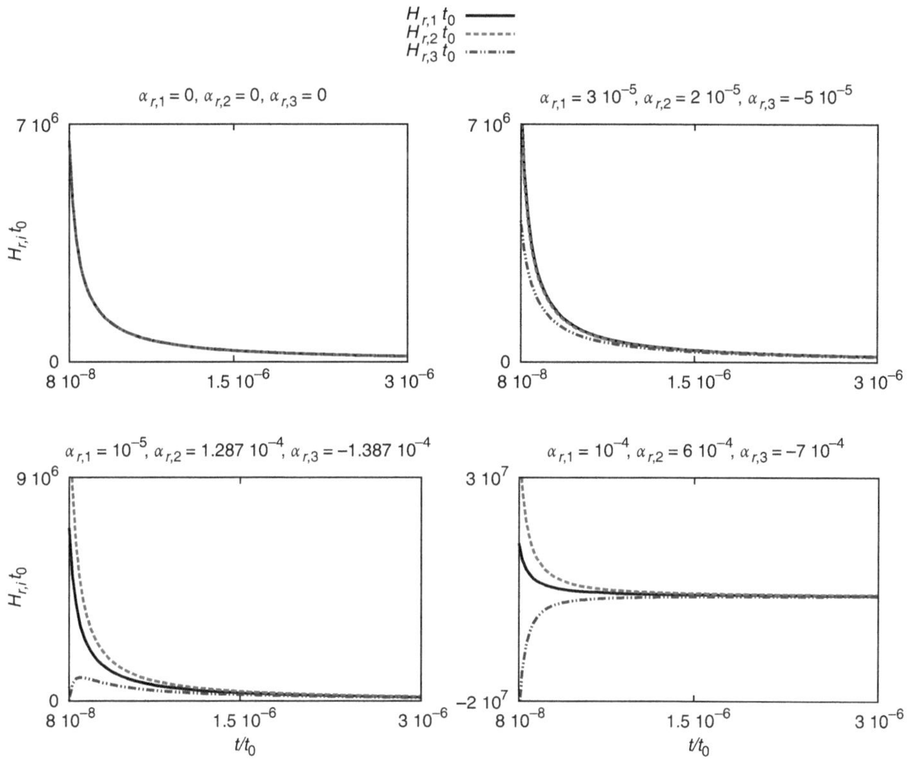

Fig. 3.2: Evolution of the normalized directional Hubble parameters in the radiation dominated BI model starting at $t/t_0 = 7.7 \cdot 10^{-8}$ with different sets of anisotropy coefficients $\alpha_{r,i}$.

which gives the total volume of the FRW one:

$$\tilde{V}_r = (a_{nr,1}\ a_{nr,2}\ a_{nr,3})^3 = \left(\frac{t}{t_0}\right)^2 .$$

Here, due to the consistency relation of integration constants (3.20), the sum of the normalized anisotropy coefficients $\alpha_{r,i}$ disappears

$$K_{r,1} + K_{r,2} + K_{r,3} = 0 \Longrightarrow \alpha_{r,1} + \alpha_{r,2} + \alpha_{r,3} = 0 . \tag{3.33}$$

As a result, the total volume element is not affected by the directional initial anisotropic expansion/contraction(s) in the early phase of a BI radiation dominated universe.

4 Metric waves in a nonstationary universe and dissipative oscillator

The application of the Einstein field equations to different cosmological models produces metrics dependent on time with exponential growth and decay. This allows us to study dissipative phenomena in the cosmological framework.

4.1 Linear metric waves in flat spacetime

The solution of the nonlinear field equations comes from linearization, which is also known as the weak field approximation. This method describes the gravitational field very far from the source of gravity. The weak field approximation is based on finding the quasi-Minkowskian coordinates $x^\mu = \{x, y, z, t\}$ (where $c = 1$) such that the metric differs from its Minkowskian form by only small quantities of $h_{\mu\nu}$,

$$g_{\mu\nu} = \eta_{\mu\nu} + \epsilon h_{\mu\nu}\,, \qquad |h_{\mu\nu}| \ll 1\,. \tag{4.1}$$

Since the metric tensor $g_{\mu\nu}$ is symmetric, $h_{\mu\nu}$ becomes symmetric as well; $h_{\mu\nu} = h_{\nu\mu}$. Apart from this, the inverse of the metric in terms of $h^{\mu\nu}$ is

$$g^{\mu\nu} = \eta^{\mu\nu} - \epsilon h^{\mu\nu}\,. \tag{4.2}$$

Here $h^{\mu\nu} = \eta^{\mu\alpha}\eta^{\nu\beta}h_{\alpha\beta}$, and the metric tensor is formed by subtracting the correction term ϵ from the Minkowski metric. Using the metric tensor in the weak field approximation (4.2), one can obtain the Christoffel symbols

$$\Gamma_{\alpha\beta}{}^{\mu} = \frac{1}{2}\left(h^{\mu}{}_{\alpha,\beta} + h^{\mu}{}_{\beta,\alpha} - h_{\alpha\beta,}{}^{\mu}\right), \tag{4.3}$$

and the Riemann tensor

$$R^{\alpha}{}_{\beta\mu\nu} = \Gamma^{\alpha}{}_{\beta\nu,\mu} - \Gamma^{\alpha}{}_{\beta\mu,\nu}\,. \tag{4.4}$$

Note that in the weak field approximation, the Riemann tensor will be contributed to only from the derivatives of Γ, not the Γ^2 terms. Then the Riemann tensor becomes

$$2R_{\alpha\beta\mu\nu} = h_{\nu\alpha,\beta\mu} + h_{\beta\mu,\alpha\nu} - h_{\beta\nu,\alpha\mu} - h_{\mu\alpha,\beta\nu}\,. \tag{4.5}$$

By raising the α into (4.5), it then becomes

$$2R^{\alpha}{}_{\beta\mu\nu} = h^{\alpha}{}_{\nu,\beta\mu} + h_{\beta\mu,\nu}{}^{\alpha} - h_{\beta\nu,\mu}{}^{\alpha} - h^{\alpha}{}_{\mu,\beta\nu}\,. \tag{4.6}$$

The contraction of the Riemann tensor (4.6) then gives the the Ricci tensor

$$2R_{\mu\nu} = h^{\alpha}{}_{\nu,\mu\alpha} + h_{\mu\alpha,\nu}{}^{\alpha} - h_{\mu\nu,\alpha}{}^{\alpha} - h^{\alpha}{}_{\alpha,\mu\nu}\,. \tag{4.7}$$

https://doi.org/10.1515/9783110515367-004

Here, it is possible to rewrite the Ricci tensor as follows:

$$2R_{\mu\nu} = h^{\alpha}{}_{\nu,\mu\alpha} + h_{\mu\alpha,\nu}{}^{\alpha} - h_{\mu\nu,\alpha}{}^{\alpha} - h_{,\mu\nu} \tag{4.8}$$

taking into account the definition $h \equiv h^{\alpha}{}_{\alpha}$. Following this, one can obtain the Ricci scalar by raising index μ and assuming $\mu = \nu$ and $\alpha = \beta$ in the Ricci tensor (4.8)

$$R = h_{\mu\beta,}{}^{\mu\beta} - h_{,\beta}{}^{\beta} \,. \tag{4.9}$$

Then, substituting the Ricci tensor (4.8), the Ricci scalar (4.9) and the metric tensor (4.2) into the Einstein field equation, we obtain the weak field approximation to the field equations:

$$h^{\alpha}{}_{\nu,\alpha}{}^{\mu} + h^{\mu}{}_{\alpha,\nu}{}^{\alpha} - h^{\mu}{}_{\nu,\alpha}{}^{\alpha} - h_{,\nu}{}^{\mu} - \eta_{\mu\nu}\left(h_{\mu\beta,}{}^{\mu\beta} - h_{,\beta}{}^{\beta}\right) = \frac{16}{c^4} T_{\mu\nu} \,. \tag{4.10}$$

Note that the field equations (4.10) are invariant under the gage transformation

$$h_{\mu\nu} \to h_{\mu\nu} + \psi_{\mu,\nu} + \psi_{\nu,\mu} \,. \tag{4.11}$$

This leads to

$$\partial^{\mu}\left(h_{\mu\nu} - \frac{1}{2}\eta_{\mu\nu} h\right) = 0 \,, \tag{4.12}$$

which is known as the Hilbert condition. Then the field equation (4.10) is reduced to

$$\partial^{\alpha}\partial_{\alpha}\left(h_{\mu\nu} - \frac{1}{2}\eta_{\mu\nu} h\right) = -\frac{16\pi G}{c^4} T_{\mu\nu} \,. \tag{4.13}$$

This expression may be written as

$$c_{\mu\nu} := h_{\mu\nu} - \frac{1}{2}\eta_{\mu\nu} h \,. \tag{4.14}$$

The equation (4.14) is the trace-reversed version of the hs. It is possible to show that $c = -h$ by using $c := c_{\lambda}{}^{\lambda}$,

$$c_{\mu\nu} := h_{\mu\nu} - \frac{1}{2}\eta_{\mu\nu} h \quad \to h_{\mu\nu} = c_{\mu\nu} - \frac{1}{2}\eta_{\mu\nu} c \,. \tag{4.15}$$

Then the Ricci tensor (4.8) can be rearranged with respect to c using the relation (4.15) and $c = -h$,

$$2R_{\mu\nu} = c_{\mu\alpha\,\nu}{}^{\alpha} + c^{\alpha}_{\nu\alpha\,\mu} - \Box h_{\mu\nu} \,, \tag{4.16}$$

in which the box is the four-dimensional Laplace operator (also called the d'Alembert operator),

$$\Box = \frac{\partial^2}{\partial t^2} - \nabla^2 \,. \tag{4.17}$$

In the Ricci tensor the first two terms on the right-hand side can vanish under the suitable coordinate transformation

$$x'^{\mu} = x^{\mu} + f^{\mu}(x) \,. \tag{4.18}$$

Here f^μ are the functions of position that are in the order of ϵ. Under this infinitesimal coordinate transformation, the metric tensor $g_{\mu\nu}$ transforms as well:

$$g_{\mu\nu} = g'_{\mu\nu}\frac{\partial x'^\alpha}{\partial x^\mu}\frac{\partial x'^\beta}{\partial x^\nu} , \tag{4.19}$$

where,

$$x'^\beta = x^\beta + f^\beta \longrightarrow \frac{\partial x'^\beta}{\partial x^\nu} = \frac{\partial x^\beta}{\partial x^\nu} + f^\beta{}_{,\nu} = \delta^\beta{}_{,\nu} + f^\beta{}_{,\nu} . \tag{4.20}$$

This also holds for x'^α. Substituting the partial derivatives of coordinates (4.20) into the metric tensor (4.19) in terms of the weak field approximation, one can obtain

$$h'_{\mu\nu} = h_{\mu\nu} - f_{\mu,\nu} - f_{\nu,\mu} . \tag{4.21}$$

If the above equation is multiplied by the Minkowski metric $\eta^{\mu\nu}$, the following relation is obtained:

$$h - h' = 2f_{\lambda,}{}^{\lambda} = c - c' . \tag{4.22}$$

To construct c forms, the h form of equations (4.15) is used in (4.22). This then leads to

$$c'_{\mu\nu} = c_{\mu\nu} + \eta_{\mu\nu} f_\lambda{}^\lambda - f_{\mu,\nu} - f_{\nu,\mu} . \tag{4.23}$$

If equation (4.23) is contracted two times by multiplying it with $\eta^{\mu\nu}$, and the derivative is taken in terms of ν, then the following equations is formed:

$$c'^{\mu\nu}_{,\nu} = c^{\mu\nu}_{,\nu} - f^{\mu\nu}{}_{,\nu} \quad . \tag{4.24}$$

Assuming that $c'^{\mu\nu}_{,\nu} = 0$, then $f^{\mu\nu}{}_{,\nu} = \Box f^\mu = c^{\mu\nu}_{,\nu}$. This assumption leads to the Ricci tensor (4.16) reduced to the following simple form:

$$2R_{\mu\nu} = -\Box h_{\mu\nu} . \tag{4.25}$$

As a result, the vacuum field equations become

$$\Box h_{\mu\nu} = 0 \ \left(c^{\mu\nu}_{,\nu} = 0\right) . \tag{4.26}$$

This partial differential equations suggest the existence of gravitational waves. In [2] was shown that the field $h^{\mu\nu}$ can be decomposed in partial waves

$$h^{\mu\nu} = \sum_\lambda \frac{1}{2\pi^{\frac{3}{2}}} \int d^3k e^{\mu\nu}{}_{k\lambda}\{u_{k\lambda}(t)e^{i\mathbf{k}x} + u^\dagger{}_{k\lambda}(t)e^{-i\mathbf{k}x}\} , \tag{4.27}$$

where $k \equiv \{k_0 = \omega = ck, \mathbf{k}\}$. The wave functions $u_{k\lambda}(t)$ satisfy the simple harmonic oscillator equation with a constant frequency,

$$\frac{d^2}{dt^2}u_{k\lambda}(t) + \omega^2 u_{k\lambda}(t) = 0 . \tag{4.28}$$

As is seen, for flat spacetime (Minkowski), the frequency of the harmonic oscillator is time independent. In the following section, gravitational waves in an expanding background will be discussed.

4.2 Metric waves in an expanding universe

The studies [2, 3] and [4] showed that in a four-dimensional spacetime, the weak approximation is applied to a flat background metric for de Sitter space

$$g^0{}_{\mu\nu} = \begin{pmatrix} 1 & 0 & 0 & 0 \\ 0 & -a^2(t) & 0 & 0 \\ 0 & 0 & -a^2(t) & 0 \\ 0 & 0 & 0 & -a^2(t) \end{pmatrix},$$

where the scale factor gives exponential growth

$$a(t) = a_0 e^{\frac{1}{3}Ht}\,.$$

Then, similar to the Euclidean metric, the expanding background gives the wave equation in which the field $h^{\mu\nu}$ is then decomposed in partial waves, as in the previous case

$$h^{\mu\nu} = \sum_{\lambda} \frac{1}{(2\pi)^{\frac{3}{2}}} \int d^3k e^{\mu\nu}{}_{k\lambda}\{u_{k\lambda}(t)e^{i\mathbf{k}x} + u^{\dagger}{}_{k\lambda}(t)e^{-i\mathbf{k}x}\}\,.$$

The main difference between Euclidean and expanding metrics comes from the wave functions $u_{k\lambda}(t)$. In the expanding universe the the wave functions satisfy the damped harmonic oscillator equation with time dependent frequency

$$\ddot{u}(t) + H\dot{u}(t) + \omega^2(t)u(t) = 0, \tag{4.29}$$

in which the time dependent frequency is

$$\omega^2(t) = \frac{k^3c^2}{a^2(t)}\,. \tag{4.30}$$

4.2.1 Hyperbolic geometry of the damped oscillator and double universe

The quantization of the one-dimensional damped harmonic oscillator with constant frequency was studied by [14], and later in [8]. It shows that the Lagrangian of (2+1)-dimensional pseudo-Euclidean topologically massive electrodynamics presents the same form as the Lagrangian of the damped harmonic oscillator. A paper by [4] stresses the problem of quantization of the inflationary time evolution of an expanding universe due to the problem of quantizing the nonunitary time evolution dynamics. Following this, [3] and [2] studied the quantization of the expanding metrics. In these studies they show that the canonical quantization can be successfully applied to the expanding metrics by doubling the degrees of freedom of the system. This doubling of the degrees of freedom is required in the study of dissipation, which provides a closed system in which to perform the canonical quantization.

Here, using the damping oscillator equation (4.31) that is derived by applying the weak approximation to the de Sitter expanding metric, one can obtain a double oscillator system to perform the canonical quantization in:

$$\ddot{u}(t) + H\dot{u}(t) + \omega^2(t)u(t) = 0 , \tag{4.31}$$

$$\ddot{v}(t) - H\dot{v}(t) + \omega^2(t)v(t) = 0 . \tag{4.32}$$

A study by [4] describes this double oscillatory system, which is produced by the expanding metric, as the double universe. Equation (4.31) is generated from the inflating metric while equation (4.32) is obtained from the deflating metric. As is seen, this set of equations form a closed system. Therefore, we can obtain the Lagrangian of this closed system/double universe in terms of wave functions u and v

$$L = m\dot{u}\dot{v} + \frac{H}{2}(u\dot{v} - \dot{u}v) - \omega^2(t)uv . \tag{4.33}$$

Using the Lagrangian (4.33), the canonical momenta for each mode u and v are obtained:

$$p_u = \frac{\partial L}{\partial \dot{u}} = m\dot{v} - \frac{H}{2}v , \tag{4.34}$$

and

$$p_v = \frac{\partial L}{\partial \dot{v}} = m\dot{u} + \frac{H}{2}u . \tag{4.35}$$

As a result, one can find the Hamiltonian of the system by using the momenta and the Lagrangian

$$\mathcal{H} = p_u\dot{u} + p_v\dot{v} - L = p_u p_v + \frac{1}{2}H(vp_v - up_u) + \Omega^2 uv , \tag{4.36}$$

in which $\Omega^2(t)$ is the time dependent frequency of the double universe,

$$\Omega^2(t) \equiv \left(\omega^2(t) - \frac{H^2}{4}\right) , \tag{4.37}$$

for which we will get a real $\Omega > 0$ for an inflating universe. The reason is implied by $\omega^2(t) > \frac{H^2}{4}$ and then $0 \leq t < \frac{3}{H}\ln\left(\frac{2k^3c}{a_0H}\right)$ with $\left(\frac{2k^3c}{a_0H}\right) \geq 1$. There is also an interesting relation between u and v:

$$v = ue^{Ht} . \tag{4.38}$$

If one substitutes

$$u(t) = \frac{1}{\sqrt{2}}r(t)e^{\frac{-Ht}{2}} , \quad v(t) = \frac{1}{\sqrt{2}}r(t)e^{\frac{Ht}{2}}$$

into the system of equations (4.31) and (4.32), finally the doublet damping oscillator system is reduced to a simple oscillator form

$$\ddot{r} + \Omega^2(t)r = 0 , \tag{4.39}$$

which is also known as the parametric oscillator in terms of $r(t)$. This clarifies the meaning of the doubling oscillator, since the $u - v$ oscillator is a noninflating or nondeflating system together. In a way, the system is closed and the energy is conserved.

This is why it is now possible to set up the canonical quantization. To obtain the quantization of the system, the following commutators are introduced:

$$[u, p_u] = i\hbar = [v, p_v] , \quad [u, v] = 0 = [p_u, p_v] . \tag{4.40}$$

These commutations hold for the new variables as well. The new variables U and V are given by the following transformations:

$$U(t) = \frac{u(t) + v(t)}{\sqrt{2}} \tag{4.41}$$

and

$$V(t) = \frac{u(t) - v(t)}{\sqrt{2}} . \tag{4.42}$$

These transformations ($u \to U$ and $v \to V$) are the hyperbolic transformations, and they preserve the commutation relation as aforementioned:

$$[U, p_U] = i\hbar = [V, p_V] , \quad [U, V] = 0 = [p_U, p_V] . \tag{4.43}$$

Now the parametric oscillator of $r(t)$ can be decomposed on the hyperbolic plane in terms of the new variables U and V. As a result, the parametric oscillator becomes

$$r^2(t) = U^2(t) - V^2(t) . \tag{4.44}$$

The Lagrangian (4.33) of the double system changes its form in terms of U and V

$$L = L_{0,U} - L_{0,V} + \frac{H}{2}(\dot{U}V - \dot{V}U) , \tag{4.45}$$

where $L_{0,U}$ and $L_{0,V}$ are

$$L_{0,U} = \frac{1}{2}\dot{U}^2 - \frac{\omega^2(t)}{2}U^2 \tag{4.46}$$

and

$$L_{0,V} = \frac{1}{2}\dot{V}^2 - \frac{\omega^2(t)}{2}V^2 . \tag{4.47}$$

Equation (4.45) shows that the dissipation or inflation term H is the Hubble constant, which acts as a coupling between the oscillators U and V. Therefore, the Hubble constant produces a correction to the kinetic energy for both oscillators. Then the momenta of the new parametric oscillator are obtained:

$$p_U = \dot{U} + \frac{H}{2}V \tag{4.48}$$

and

$$p_V = -\dot{V} - \frac{H}{2}U , \tag{4.49}$$

and the Hamiltonian of the system becomes

$$\mathcal{H} = \mathcal{H}_1 - \mathcal{H}_2 = \frac{1}{2}\left(p_U - \frac{H}{2}V\right)^2 + \frac{\omega^2(t)}{2}U^2 - \frac{1}{2}\left(p_V + \frac{H}{2}U\right)^2 - \frac{\omega^2(t)}{2}V^2 . \tag{4.50}$$

Finally, the system is quantized using this Hamiltonian, which is in hyperbolic form.

5 Bosonic and fermionic models of a Friedman–Robertson–Walker universe

Taking into account that the Friedman equations can be simplified into a harmonic oscillator differential equation with a constant frequency, [37] applied the factorization procedure to the solution of the oscillator equation, which was classified into to two classes: bosonic (nonsingular) and fermionic (singular). These simple solutions of the oscillator equation in two different geometries (open and closed) are defined as cosmological zero modes by [36]. In this chapter, we briefly review these differential equations in a mathematical scheme.

As mentioned, FRW cosmologies in comoving time obey the Friedman equations

$$\frac{\ddot{a}}{a} = -\frac{4\pi G}{3}(\rho + 3p) , \tag{5.1}$$

$$H_0{}^2 \equiv \left(\frac{\dot{a}}{a}\right)^2 = \frac{8\pi G\rho}{3} - \frac{\kappa}{a^2} , \tag{5.2}$$

$$p = (\gamma - 1)\rho , \tag{5.3}$$

where γ is related to the adiabatic parameter $\omega = \gamma - 1$ of the cosmological fluid, which has been described in detail. In addition, here the velocity of light is assumed to be $c = 1$.

5.1 Bosonic Friedman–Robertson–Walker cosmology

Moving on to the conformal time variable η, defined through $dt = a(\eta)d\eta$, we can combine the three equations in a Riccati equation for the Hubble parameter H_η as follows:

$$H_0{}'(\eta) + cH_0^2(\eta) + c\kappa = 0 . \tag{5.4}$$

Proof. Let (5.3) substitute into (5.1), so that equation (5.1) becomes

$$\frac{\ddot{a}}{a} = -\frac{4\pi G}{3}\left[(3\gamma - 2)\rho\right] . \tag{5.5}$$

By using the equation (5.5), we get the density ρ as

$$\rho = -\frac{3}{4\pi G}\frac{\ddot{a}}{a}\frac{1}{(3\gamma - 2)} . \tag{5.6}$$

Insert (5.6) ρ into (5.2) so that

$$\left(\frac{\dot{a}}{a}\right)^2 = \left[-2\frac{\ddot{a}}{a}\frac{1}{(3\gamma - 2)}\right] - \frac{\kappa}{a^2} . \tag{5.7}$$

https://doi.org/10.1515/9783110515367-005

Arranging equation (5.7), we obtain

$$\left(\frac{3}{2}\gamma - 1\right)\left(\frac{\dot{a}}{a}\right)^2 = -\frac{\ddot{a}}{a} - \left(\frac{3}{2}\gamma - 1\right)\frac{\kappa}{a^2} . \tag{5.8}$$

For simplification in notation, let us assume that $\beta \equiv (\frac{3}{2}\gamma - 1)$. Hence, equation (5.8) is reduced to the following form:

$$-\frac{\ddot{a}}{a} = \beta\left(\frac{\dot{a}}{a}\right)^2 + \beta\frac{\kappa}{a^2} . \tag{5.9}$$

Multiply equation (5.9) by a^2, so that

$$-\ddot{a}a = \beta\dot{a}^2 + c\kappa . \tag{5.10}$$

Using conformal time, we can then construct equation (5.10) in terms of the Hubble parameter. To write equation (5.10) in terms of the Hubble parameter, we need to obtain the first and the second derivatives of the scale factor $a(t)$ with respect to the conformal time as follows:

$$dt = a(\eta)d\eta \longrightarrow \frac{dt}{d\eta} = a(\eta) . \tag{5.11}$$

Using the chain rule, we get

$$\frac{da}{d\eta} = \frac{da}{dt}\frac{dt}{d\eta} \longrightarrow \acute{a} = \dot{a}a(\eta) \longrightarrow \frac{a'}{a} = \dot{a} . \tag{5.12}$$

The second derivative in terms of η is given by

$$\frac{d^2a}{d\eta^2} = \frac{d^2a}{dt^2}\left(\frac{dt}{d\eta}\right)^2 + \frac{da}{dt}\frac{d^2t}{d\eta^2} \longrightarrow a'' = \ddot{a}a^2 + \dot{a}a' . \tag{5.13}$$

In light of equations (5.12) and (5.13), the second derivative of $a(t)$ in terms of time can be written as

$$\ddot{a} = \frac{a''}{a^2} - \frac{a''^2}{a^3} . \tag{5.14}$$

Substituting equations (5.14) and (5.12) in (5.10) gives the following equation:

$$\beta\left(\frac{a'}{a}\right)^2 + \beta\kappa = -\left[\frac{a''}{a} - \frac{a'^2}{a^2}\right] . \tag{5.15}$$

From this, it is clear that $H_0(\eta) = \frac{a'}{a}$. Therefore, equation (5.15) becomes

$$H_0{}'(\eta) + \beta H_0^2(\eta) + \beta\kappa = 0 .$$ □

Equation (5.4) is known as the Riccati equation. To linearize this Riccati equation, the Cole–Hopf transformation should be given as $H_0(\eta) = \frac{1}{\beta}\frac{\Theta'}{\Theta}$. After substituting this transformation, the linearized equation is given as

$$\frac{1}{\beta}\frac{\Theta''}{\Theta} + \beta\kappa = 0 . \tag{5.16}$$

Arrange (5.16) as follows:

$$\Theta'' - \beta \times \beta_{\kappa,b}\Theta = 0\,, \tag{5.17}$$

where $\beta_{\kappa,b} = -\kappa\beta$. Additionally, the particular Riccati solutions for the positive and negative curvature indices are discussed as follows [37]:

- Case $\kappa = 1$: The positive curvature index means that the constant $\beta_{\kappa,b}$ becomes

$$\beta_{\kappa,b} = -1 \cdot \beta = -\beta \tag{5.18}$$

Therefore, equation (5.17) becomes

$$\Theta''_{1,b} + \beta^2\Theta_{1,b} = 0\,, \tag{5.19}$$

which is the standard oscillator equation. The solution of this second-order differential equation is obtained as

$$\Theta_{1,b}(\eta) = A_1 \sin(\beta\eta) + A_2 \cos(\beta\eta)\,, \tag{5.20}$$

where A_1 and A_2 are the constants, which can be obtained only by the initial conditions. As can be seen, the solution of the oscillator equation for the bosonic case in a closed geometry corresponds to an overdamped harmonic oscillator. The solution of (5.20) can be reduced to the following equation by adding an arbitrary phase factor ϕ:

$$\Theta_{1,b}(\eta) = A\cos(\beta\eta + \phi)\,, \tag{5.21}$$

where A is the integration constant. If we substitute (5.21) and its first derivatives in the Hubble parameter, the Hubble parameter becomes

$$H_0{}^+(\eta) - \frac{1}{\beta}\frac{\Theta'_{1,b}}{\Theta_{1,b}} = -\left(\frac{\sin(\beta\eta + \phi)}{\cos(\beta\eta + \phi)}\right) = -\tan(\beta\eta + \phi)\,. \tag{5.22}$$

Equation (5.22) is the particular solution of the Riccati equation for a closed geometry with the curvature $\kappa = 1$. This solution also indicates that the Hubble parameter presents a periodic motion as oscillations. Figure 5.1 shows the evolution of the Hubble parameter in three different eras: radiation dominated with $\gamma = 4/3$, dark matter dominated with $\gamma = 1$, and dark energy dominated with $\gamma = 0$, derived for bosons in the closed geometry. As is seen from the figure, Hubble parameters in these three different eras oscillate and accept negative values as well as positive ones. While negative values of the Hubble parameter indicate the contraction of the bosonic universe, the positive values show expansion of the universe.

- Case $\kappa = -1$: For negative curvature, the constant $c_{\kappa,b}$ is transformed into the following:

$$\beta_{\kappa,b} = -(-1) \cdot \beta = \beta\,. \tag{5.23}$$

Hence, (5.17) becomes

$$\Theta''_{-1,b} - \beta^2\Theta_{-1,b} = 0\,. \tag{5.24}$$

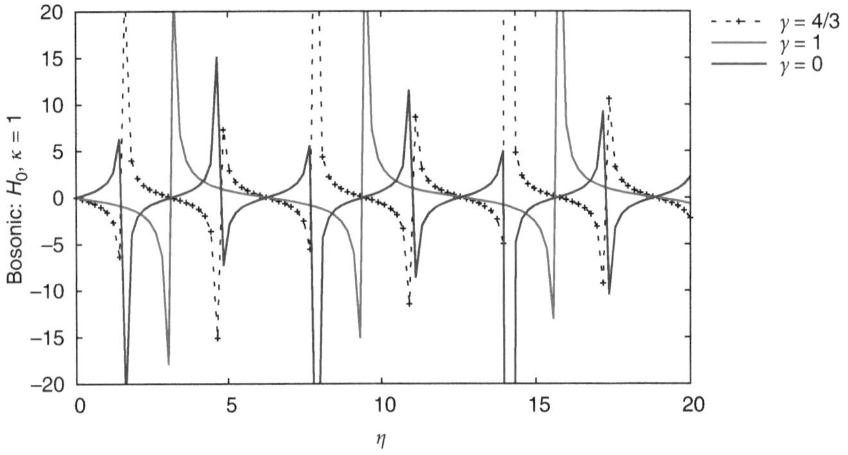

Fig. 5.1: Evolution of the bosonic Hubble parameter, H_0, in terms of η for closed geometry ($\kappa = 1$) in three different epochs: radiation dominated with $\gamma = 4/3$, dark matter dominated with $\gamma = 1$, and dark energy dominated with $\gamma = 0$. In these three cases, the phase is assumed to be zero, $\phi = 0$.

The solution of this differential equation is then

$$\Theta_{-1,b}(\eta) = B_1 \sinh \beta\eta + B_2 \cosh \beta\eta \,. \tag{5.25}$$

This solution corresponds to an underdamped harmonic oscillator satisfied by equation (5.24). In addition to this, equation (5.25) can be written as

$$\Theta_{-1,b}(\eta) = B \sinh c\eta \,. \tag{5.26}$$

This solution of the scale factor in terms of conformal time in open geometry leads us to the Hubble parameter

$$H_0{}^{-}(\eta) = \frac{1}{\beta}\frac{\Theta'_{-1,b}}{\Theta} = \frac{\cosh \beta\eta}{\sinh \beta\eta} = \coth \beta\eta \,, \tag{5.27}$$

in which the symbol (−) means negative curvature. Equation (5.27) is the particular solution of the Riccati equation for $\kappa = -1$. This indicates that the Hubble parameter of bosonic cosmology in open geometry shows a hyperbolic-type motion as under damped oscillations. Figure 5.2 presents the evolution of the bosonic Hubble parameter in terms of η in an open geometry for three different epochs. The figure indicates that the three Hubble parameters in this setting show overdamping behavior. While the Hubble parameter of the dark matter dominated model reaches the constant phase after expansion, the dark energy dominated universe for the bosonic model reaches a stable condition after contraction.

Generally, the two Hubble parameters are obtained as $H_0{}^{-}(\eta) = \coth \beta\eta$ and $H_0{}^{+}(\eta) = -\tan(c\eta + \phi)$ for the baryonic cosmology, which completely depends on the geometry of the universe. Taking into account that the bosonic Hubble parameters $H_0{}^{-}(\eta) =$

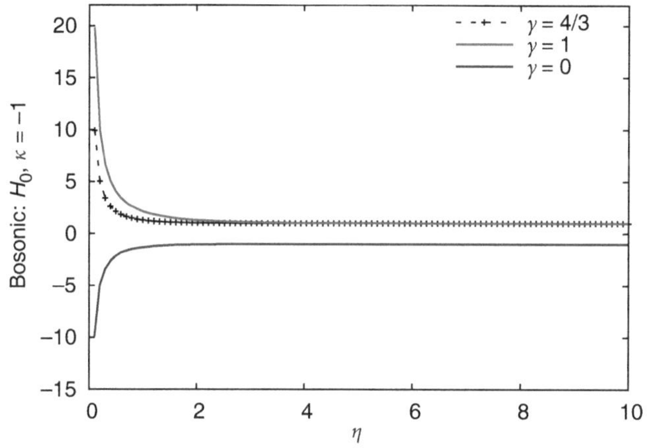

Fig. 5.2: Evolution of the bosonic Hubble parameter, H_0, in terms of η for hyperbolic geometry ($\kappa = -1$) in three different epochs: radiation dominated with $\gamma = 4/3$, dark matter dominated with $\gamma = 1$, and dark energy dominated with $\gamma = 0$.

$\coth\beta\eta$ and $H_0{}^+(\eta) = -\tan(\beta\eta + \phi)$ are related to the common factorizations of equation (5.17), the equation can be rearranged as follows:

$$\left(\frac{d}{d\eta} + \beta H_0\right)\left(\frac{d}{d\eta} - \beta H_0\right)\Theta = 0\,, \tag{5.28}$$

and

$$\left(\frac{d^2\Theta}{d\eta^2} - \beta\frac{dH_0}{d\eta}\Theta - \beta^2 H_0^2\Theta\right) = 0\,. \tag{5.29}$$

Using (5.29), we obtain the following second-order differential equation for the bosonic case:

$$\Theta'' - \beta(H_0' + \beta H_0^2)\Theta = 0\,. \tag{5.30}$$

Equations (5.30) and (5.28) are equivalent. We can combine them as follows:

$$\left(\frac{d}{d\eta} + \beta H_0\right)\left(\frac{d}{d\eta} - \beta H_0\right)\Theta = \Theta'' - \beta(H_0' + \beta H_0^2)\Theta = 0\,. \tag{5.31}$$

Borrowing a term from supersymmetric quantum mechanics, we call the solutions Θ bosonic zero modes in terms of scale factors. Firstly, the Hubble parameter is defined as

$$H = \frac{1}{\beta}\frac{d\Theta}{\Theta} = \frac{da}{a}\,. \tag{5.32}$$

The solution of this first-order differential equation is

$$\Theta^{\frac{1}{\beta}} = Ca\,, \tag{5.33}$$

in which C is integration constant. Generally, we can write

$$\Theta^{\frac{1}{\beta}} \approx a\,. \tag{5.34}$$

This result can be specified for the closed and open geometries as follows:

– Case $\kappa = 1$

$$\Theta^{\frac{1}{\beta}}{}_{1,b} \approx a_{1,b} \tag{5.35}$$

and

$$\Theta_{1,b}(\eta) \approx \cos(\beta\eta + \phi) \longrightarrow a_{1,b} \approx [\cos(\beta\eta + \phi)]^{\frac{1}{\beta}} . \tag{5.36}$$

Figure 5.3 shows the evolution of the scale factors of the bosonic zero modes in closed geometry in three different universe models. As seen, in closed geometry the scale factors show oscillatory behaviors. It is crucial to note that the dark energy dominated model with $\gamma = 0$ does not provide a physically accepted result, due to the fact that its scale factor accepts negative values as well as positive values.

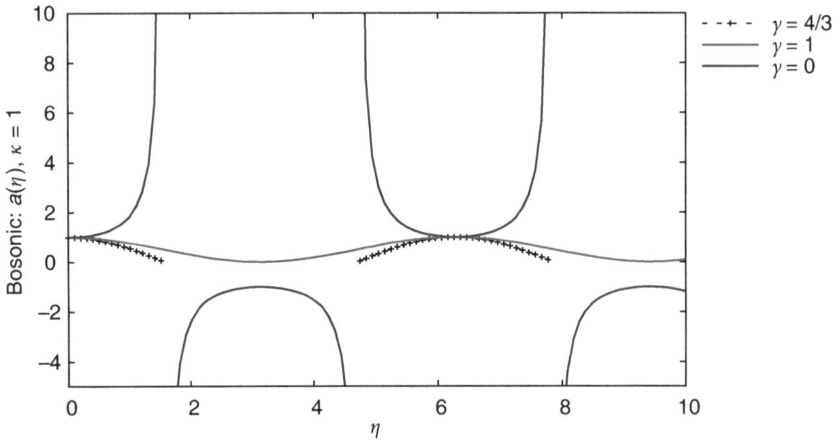

Fig. 5.3: The scale factors of the bosonic models in three different eras (radiation dominated with $\gamma = 4/3$, dark matter dominated with $\gamma = 1$, and dark energy dominated with $\gamma = 0$) in a closed geometry, based on equation (5.36).

– Case $\kappa = -1$

$$\Theta^{\frac{1}{\beta}}_{-1,b} \approx a_{-1,b} \tag{5.37}$$

$$\Theta_{-1,b}(\eta) \approx \sinh\beta\eta \longrightarrow a_{-1,b} \approx [\sinh\beta\eta]^{\frac{1}{\beta}} \tag{5.38}$$

Finally, we can conclude that a universe with open geometry has a scale factor showing underdamped oscillator characteristics, just as in the contracting and expanding universe for bosonic zero modes. As is seen from Figure 5.4, both dark matter and radiation dominated open models show expansion. The dark energy dominated open universe presents a negative scale factor, which approaches zero for high η values. It is important to note that in our universe, a negative scale factor

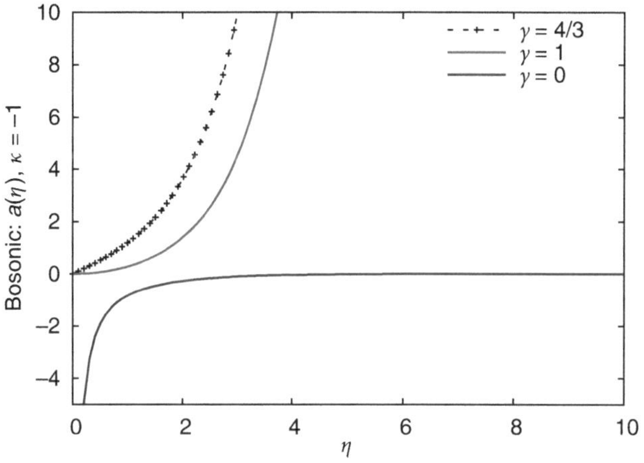

Fig. 5.4: The scale factors of the open bosonic models in three different eras (radiation dominated with $\gamma = 4/3$, dark matter dominated with $\gamma = 1$, and dark energy dominated with $\gamma = 0$) based on equation (5.38).

is not observed. Therefore the dark energy solution of the scale factor of the open bosonic model does not provide a physically acceptable condition.

5.2 Fermionic Friedman–Robertson–Walker cosmology

A class of barotropic FRW cosmologies, with inverse scale factors with respect to the bosonic ones, can be obtained by considering the supersymmetric partner; in other words, a fermionic form of equation (5.30), which is obtained by applying the factorization brackets in reverse order:

$$\left(\frac{d}{d\eta} - \beta H_0\right)\left(\frac{d}{d\eta} + \beta H_0\right)\Theta = 0 \tag{5.39}$$

and

$$\left(\frac{d^2\Theta}{d\eta^2} + \beta\frac{dH_0}{d\eta}\Theta - \beta^2 H_0^2\Theta\right) = 0\,. \tag{5.40}$$

We can write equation (5.40) as

$$\Theta'' - \beta(-H_0' + \beta H_0^2)\Theta = 0\,. \tag{5.41}$$

Equation (5.41) can be rewritten as

$$\Theta'' - \beta.\beta_{\kappa,f}\Theta = 0\,, \tag{5.42}$$

where

$$\beta_{\kappa,f}(\eta) = -H_0' + \beta H_0^2 = \begin{cases} \beta(1 + 2\tan^2\beta\eta)\,, & \text{if } \kappa = 1 \\ \beta(-1 + 2\coth^2\beta\eta)\,, & \text{if } \kappa = -1 \end{cases} \tag{5.43}$$

denotes the supersymmetric partner adiabatic index of the fermionic type, associated through the mathematical scheme with the constant bosonic index. Notice that the fermionic adiabatic index is time dependent. The fermionic Θ solutions are

$$\Theta_{1,f} = \frac{\beta}{\cos(\beta\eta + \phi)} \tag{5.44}$$

and

$$\Theta_{-1,f} = \frac{\beta}{\sinh \beta\eta} \tag{5.45}$$

for $\kappa = 1$ and $\kappa = -1$ respectively. In addition, the relation between scale factor and Θ is similar to that seen before:

$$\Theta^{\frac{1}{\beta}} \approx a\,. \tag{5.46}$$

– Case $\kappa = 1$:

$$\Theta^{\frac{1}{\beta}}{}_{1,f} \approx a_{1,f}\,, \tag{5.47}$$

and

$$\Theta_{1,f}(\eta) \approx \frac{\beta}{\cos(\beta\eta + \phi)} \longrightarrow a_{1,f} \approx [\cos(\beta\eta + \phi)]^{-\frac{1}{\beta}} \tag{5.48}$$

– Case $\kappa = -1$:

$$\Theta_{-1,f}{}^{\frac{1}{\beta}} \approx a_{-1,f}\,, \tag{5.49}$$

and

$$\Theta_{-1,f}(\eta) \approx \sinh \beta\eta \longrightarrow a_{-1,f} \approx [\sinh \beta\eta]^{-\frac{1}{\beta}} \tag{5.50}$$

The evolution of the scale factors of Fermionic models for closed and open geometries in terms of η is shown in Figure 5.5. As is seen in the upper panel of Figure 5.5, similar to the dynamical behavior of the bosonic solutions in the closed geometry, the Fermionic scale factors in the closed geometry present oscillatory characteristics. In this case, the dark matter (solid red line) and the radiation (dotted black line) dominated bosonic models show initial expansion at $\eta = 0$, which later turns into contraction and discrete expansion. The dark energy dominated fermionic closed model (solid blue line) oscillates smoothly, indicating continuous transition between contraction and expansion.

If we look at the open fermionic model in terms of different cosmological constitutes, we see different dynamical characteristics for the scale factors compared to the ones given by the closed model. The lower panel in Figure 5.5 shows overdamping behavior of the scale factors for the radiation and the dark matter dominated models in open geometry. However, the scale factor of the dark energy dominated model accepts negative values, which is not seen in the observed universe.

Another important point to mention before finalizing this chapter is that the bosonic,

$$a_{1,b} \approx [\cos(\beta\eta + \phi)]^{\frac{1}{\beta}}\,, \tag{5.51}$$

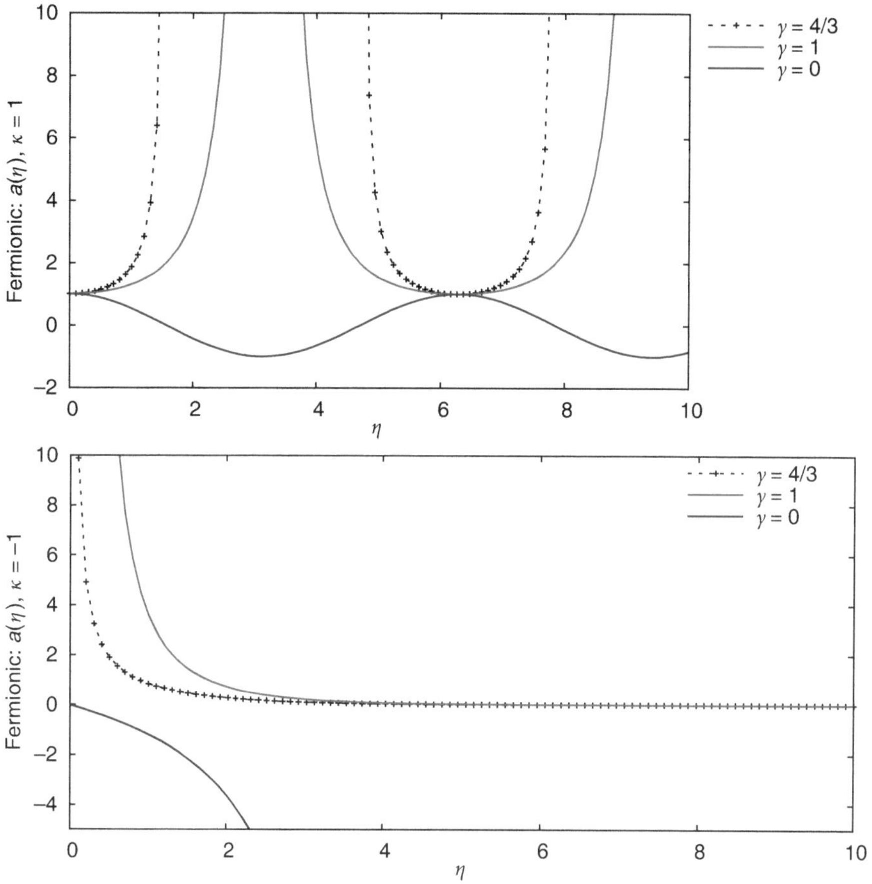

Fig. 5.5: Evolution of the scale factors of the closed (upper) and open (lower) fermionic models in three different eras (radiation dominated with $\gamma = 4/3$, dark matter dominated with $\gamma = 1$, and dark energy dominated with $\gamma = 0$) in terms of η.

and fermionic,

$$a_{1,f} \approx [\cos(\beta\eta + \phi)]^{-\frac{1}{c}}\,, \tag{5.52}$$

cosmological models are reciprocal to each other for the same geometrical setting. This can be formulated as

$$a_{1,b}a_{1,f} = a_{+b}a_{+f} = [\cos(\beta\eta + d)]^{\frac{1}{\beta}}[\cos(\beta\eta + \phi)]^{-\frac{1}{\beta}} = 1 \tag{5.53}$$

for the closed geometry. For negative curvature, the relations between the scale factors of bosonic and fermionic cases are given in the following equations:

$$a_{-1,b} = a_{-b} \approx [\sinh\beta\eta]^{\frac{1}{\beta}} \tag{5.54}$$

and

$$a_{-1,f} = a_{-f} \approx [\sinh\beta\eta]^{-\frac{1}{\beta}}\,. \tag{5.55}$$

Again, the multiplication of these two quantity gives the same result

$$a_{-1,b}a_{-1,f} = a_{-b}a_{-f} = [\sinh\beta\eta]^{\frac{1}{\beta}}[\sinh\beta\eta]^{-\frac{1}{\beta}} = 1\,. \tag{5.56}$$

Then, we can write these results in general as

$$a_{\pm b}a_{\pm f} = cst\,. \tag{5.57}$$

Thus, bosonic expansion corresponds to fermionic contraction, and bosonic contraction becomes fermionic expansion (Figure 5.6).

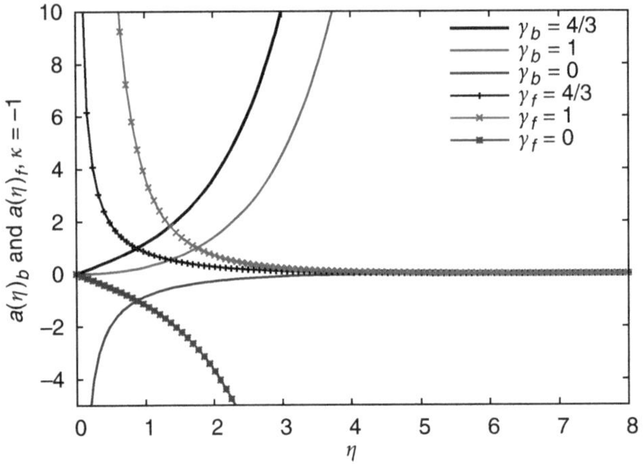

Fig. 5.6: The reciprocal nature of bosonic (with subindex b) and fermionic (with subindex f) cosmologies in terms of η in three different eras (radiation dominated with $\gamma = 4/3$, dark matter dominated with $\gamma = 1$, and dark energy dominated with $\gamma = 0$).

Finally, a very important question can arise in the case of the decoupled fermionic and bosonic FRW models: Is it possible to observe an oscillatory behavior for the decoupled fermionic and bosonic FRW models? To answer this question, [17] investigated the Dirac equation in the supersymmetric nonrelativistic formalism. According to this study, it is shown that the Dirac equation with the Lorentz scalar potential is associated with a supersymmetry (SUSY) pair of Schrödinger Hamiltonians. Following this study, [37] applied the method to the bosonic and fermionic models. Then [37] showed that it was a useful exercises in the case of the decoupled 'zero-mass' in which the cosmological matrix equation is

$$\sigma_y D_\eta W + \sigma_x(i\beta H_0)W = 0\,. \tag{5.58}$$

Here W is the 2×1 matrix, $W = \begin{pmatrix}\Theta_1\\ \Theta_2\end{pmatrix}$, which is a two-component zero-mass spinor, and σ_x and σ_y are defined as the following matrices:

$$-i\sigma_y = -i\begin{pmatrix}0 & -i\\ i & 0\end{pmatrix}$$

and

$$\sigma_x = \begin{pmatrix} 0 & 1 \\ 1 & 0 \end{pmatrix} .$$

Equation (5.58) is equivalent to the following decoupled equations:

$$\left[\begin{pmatrix} 0 & -i \\ i & 0 \end{pmatrix} D_\eta + \begin{pmatrix} 0 & 1 \\ 1 & 0 \end{pmatrix} (i\beta H_0) \right] \begin{pmatrix} \Theta_1 \\ \Theta_2 \end{pmatrix} = 0 . \tag{5.59}$$

Then,

$$\left[\begin{pmatrix} 0 & i(\beta H_0 - D_\eta) \\ i(D_\eta + \beta H_0) & 0 \end{pmatrix} \right] \begin{pmatrix} \Theta_1 \\ \Theta_2 \end{pmatrix} = 0 , \tag{5.60}$$

hence from matrix multiplication, we get the system of equations

$$i\,(\beta H_0 + D_\eta)\,\Theta_1 = 0 \tag{5.61}$$

and

$$i(\beta H_0 - D_\eta)\Theta_2 = 0 . \tag{5.62}$$

As can be seen, although the solutions of these differential equations are trivial, the 'zero-mass' case does not produce oscillatory-type equations. However, these equations do show that the matrix equation contains two reciprocal barotropic cosmologies, which are the same as the two components of the spinor W.

6 Time dependent constants in an oscillatory universe

In observational cosmology, the evolution of the universe is described by Einstein's field equations together with a perfect fluid and the equation of state, all of which were discussed in a great detail in the previous chapters. As we know, the Einstein field equations include gravitational and cosmological constants. In this framework the gravitational constant G becomes a coupling constant between the geometry and the matter content of the universe. Considering the universe itself as a dynamical system, Dirac proposed the large numbers hypothesis (LNH), which considered the possibility of a cosmological model in which this coupling term, the gravitational constant, varies with time ([18]).

In addition to the cosmological models with a time dependent gravitational constant, [12] suggested that the cosmological constant may vary in terms of cosmological timescale. In addition to these models, [13] studied a spatially homogeneous and isotropic cosmological model using the equation of state $p = (\gamma - 1)\rho$, where the γ varies with cosmic time. Following this, a spatially homogeneous and isotropic FRW line element was considered with time dependent gravitational and cosmological constants by [41]. In the same study, [41] applied the gamma-law equation of state of [13], in which the parameter γ depends on the scale factor $a(t)$.

Progressing from the studies mentioned above, we here give the possible solutions and derivations of the field equations, which are oscillatory equations by their nature, by taking into account the time dependent constants of general relativity. The solutions of the field equations will be investigated in two different cosmic epochs: inflationary and radiation dominated.

6.1 Model and field equations

As we discussed, the exact solutions of the field equations are obtained by using the equation of state $p = (\gamma - 1)\rho$. Now we assume that γ varies continuously with cosmological time. The functional form of $\gamma(a)$ is used to analyze a wide range of cosmological solutions in the early universe for two phases in cosmic history by [13]:

- Inflationary phase
- Radiation dominated phase.

The corresponding physical interpretations of the cosmological solutions will be discussed in this chapter.

We here consider a spatially homogeneous, isotropic Robertson–Walker line element, where the universe is assumed to be filled with a distribution of matter represented by the energy-momentum tensor of a perfect fluid (where we choose $c = 1$ for

https://doi.org/10.1515/9783110515367-006

simplicity). In this framework, the field equations with time dependent cosmological and gravitational constants are given by [42]:

$$R_{\mu\nu} - \frac{1}{2} g_{\mu\nu} R = 8\pi G(t) T_{\mu\nu} + \Lambda(t) g_{\mu\nu} , \qquad (6.1)$$

where $G(t)$ and $\Lambda(t)$ are the time dependent gravitational and cosmological constants. Here another important equation indicating the time changes of G and Λ is derived using the covariant divergence of equation (6.1):

$$[8\pi G T_{\mu\nu} + \Lambda g_{\mu\nu}]^{;\nu} = 0 . \qquad (6.2)$$

Equations (6.1) and (6.2) should be taken as the fundamental equations of gravity in which G and Λ are nothing but the coupling parameters. Then Einstein's field equations (6.1) give two independent equations, which are called Friedman equations with the time dependent G and Λ parameters

$$\frac{2a\ddot{a} + \dot{a}^2 + \kappa}{a^2} - \Lambda(t) = -8\pi G(t) p , \qquad (6.3)$$

$$\frac{3(\dot{a} + \kappa)}{a^2} \Lambda(t) = 8\pi G(t) \rho . \qquad (6.4)$$

After rearranging the above dynamical equations, (6.3) and (6.4) become

$$3\ddot{a} = -4\pi G(t) a - \left[3p + \rho - \frac{\Lambda(t)}{4\pi G(t)} \right] , \qquad (6.5)$$

$$3\dot{a}^2 = 8\pi G(t) a^2 \left[\rho + \frac{\Lambda(t)}{8\pi G(t)} \right] - 3\kappa . \qquad (6.6)$$

Proof. Firstly, multiply equation (6.3) by three, which yields

$$3\ddot{a} = -12\pi G(t) a p - \frac{3}{2} \left(\frac{\dot{a}^2 + \Lambda a^2 + \kappa}{a} \right) , \qquad (6.7)$$

and then using the second Friedman equation (6.4) we obtain

$$\dot{a}^2 = \frac{8\pi G(t) a^2}{3} \left(\rho + \frac{\Lambda}{8\pi G(t)} \right) - \kappa . \qquad (6.8)$$

Substituting the above equation (6.8) into equation (6.5), the following is obtained:

$$3\ddot{a} = -4\pi G(t) a \left[3p + \rho - \frac{\Lambda(t)}{4\pi G(t)} \right] . \qquad (6.9)$$

Similarly, we rearrange equation (6.4) as follows:

$$3\dot{a}^2 = 8\pi G(t) a^2 \left[\rho + \frac{\Lambda(t)}{8\pi G(t)} \right] - 3\kappa . \qquad \square$$

In addition, in uniform cosmology with time dependent parameters ($G = G(t)$ and $\Lambda = \Lambda(t)$), the conservation equation (6.2) should be in the following form:

$$\dot{\Lambda} = -8\pi \dot{G} \rho . \qquad (6.10)$$

Proof. Applying the divergence to each term in equation (6.2), we obtain

$$8\pi\left[G(t)_{;\nu}T_{\mu\nu} + G(t)T_{\mu\nu;\nu}\right] + \Lambda(t)_{;\nu}g_{\mu\nu} + \Lambda(t)g_{\mu\nu;\nu} = 0\,. \tag{6.11}$$

Taking into account that $T_{\mu\nu;\nu} = g_{\mu\nu;\nu} = 0$ due to energy conservation, the equation (6.11) is reduced to the following form:

$$8\pi G(t)_{;\nu}T_{\mu\nu} + \Lambda(t)_{;\nu}g_{\mu\nu} = 0\,. \tag{6.12}$$

The time component of this equation is

$$8\pi G(t)_{;0}T_{00} = \Lambda(t)_{;0}g_{00}\,, \tag{6.13}$$

where the (00) component of the energy momentum tensor becomes

$$T_{00} = -pg_{00} + (p+\rho)u_0u_0 = \rho \tag{6.14}$$

and $g_{00} = 1$ by using the Minkowski space in the matrix form,

$$\mathbf{g}_{\mu\nu} = \begin{pmatrix} 1 & 0 & 0 & 0 \\ 0 & -1 & 0 & 0 \\ 0 & 0 & -1 & 0 \\ 0 & 0 & 0 & -1 \end{pmatrix}. \tag{6.15}$$

After substituting these into equation (6.13), it becomes

$$\dot{\Lambda} = -8\pi\dot{G}\rho\,.$$ □

The Friedman equations (6.5) and (6.6) alternatively are written as follows:

$$8\pi G(t)p = -\frac{2\ddot{a}}{a} - \frac{\dot{a}^2}{a^2} - \frac{\kappa}{a^2} \tag{6.16}$$

and

$$8\pi G(t)\rho = \frac{3\dot{a}^2}{a^2} + \frac{3\kappa}{a^2} - \Lambda(t)\,, \tag{6.17}$$

which are identical to the usual Friedman equations but with time dependent cosmological parameters. Here our goal is to formulate the Friedman equations (6.6) and (6.17) in terms of the Hubble parameter H. This will allow us to obtain exact solutions using the assumptions for the gravitational and cosmological constants, and the Hubble parameter. Therefore, if we write the above Friedman equations in terms of the Hubble parameter, they become

$$\dot{H} + H^2 = -\frac{4\pi}{3}G(t)\,(3p+\rho) + \frac{1}{3}\Lambda(t) \tag{6.18}$$

and

$$H^2 = \frac{8\pi}{3}G(t)\rho + \frac{1}{3}\Lambda(t) - \frac{\kappa}{a^2}\,. \tag{6.19}$$

Proof. If we differentiate the Hubble parameter in terms of time, it becomes

$$\dot{H} = \frac{d^2}{dt^2}\ln a = \frac{\ddot{a}}{a} - \left(\frac{\dot{a}}{a}\right)^2 . \tag{6.20}$$

Adding the first derivative of the Hubble parameter and the square of the Hubble parameter, we end up with the following nonlinear first-order differential equation:

$$\dot{H} + H^2 = \frac{\ddot{a}}{a} . \tag{6.21}$$

If we insert equation (6.21) into equation (6.5), then equation (6.5) turns into the Hubble parameter dependent form

$$\dot{H} + H^2 = -\frac{4\pi}{3}G(t)\,(3p+\rho) + \frac{1}{3}\Lambda(t) .$$

Following a similar strategy, substituting H^2 instead of $(\frac{d}{dt}\ln a)^2 = \frac{\dot{a}^2}{a^2}$ in equation (6.6), we get

$$H^2 = \frac{8\pi}{3}G(t)\rho + \frac{1}{3}\Lambda(t) - \frac{\kappa}{a^2} .$$ □

As mentioned, equations (6.18) and (6.19) can be solved using assumptions of a certain equation of state and the relation between the Hubble parameter H and gravitational constant $G(t)$, as well as the Hubble parameter H and the cosmological constant $\Lambda(t)$. The time dependent equation of state is defined by the adiabatic parameter, which varies with time. This time dependent adiabatic parameter indicates that the universe goes through a continuous transition from an inflationary phase to a radiation dominated phase, etc. [13] proposed that the functional form of γ depends on a scale factor that is time dependent,

$$\gamma(a) = \frac{4}{3}\frac{A\left(\frac{a}{a_0}\right)^2 + \frac{b}{2}\left(\frac{a}{a_0}\right)^b}{A\left(\frac{a}{a_0}\right)^2 + \left(\frac{a}{a_0}\right)^b} , \tag{6.22}$$

where A is a constant, and b is a free parameter related to the power of cosmic time, which lies $0 \le b < 1$. [41] makes a distinction between two evolutionary stages of the universe using the ratio between the reference value of the scale factor a_0 and the observed scale factor a. According to this, if $a \ll a_0$, the inflationary phase of the evolution of the universe is obtained. If $a \gg a_0$, then a cosmological model for a radiation dominated phase is obtained. Before obtaining the models in the given phases of the universe, our goals is to derive an equation that includes only the Hubble parameter and the cosmological parameters. To do this, one can substitute the standard equation of state into equation (6.18), which yields

$$\dot{H} + H^2 = -\frac{4\pi}{3}G(t)\,[3\gamma - 2]\,\rho + \frac{1}{3}\Lambda(t) . \tag{6.23}$$

If we eliminate ρ between equations (6.10) and (6.23), then we get

$$H'H + \frac{H^2}{a} = \frac{1}{3}\left(\frac{3}{2}\gamma - 1\right)\frac{G(t)\Lambda'}{G'a} + \frac{1}{3}\frac{\Lambda(t)}{a} , \tag{6.24}$$

in which prime denotes differentiation with respect to scale factor a.

Proof. To obtain equation (6.24), first the density parameter ρ is written using equation (6.10),

$$\rho = -\frac{1}{8\pi}\frac{\dot{\Lambda}}{\dot{G}} . \tag{6.25}$$

Next the above equation is inserted into equation (6.23), which becomes

$$\dot{H} + H^2 = -\frac{4\pi}{3}G(t)\,[3\gamma - 2]\left(-\frac{\dot{\Lambda}}{8\pi\dot{G}}\right) + \frac{1}{3}\Lambda(t) . \tag{6.26}$$

After rearranging equation (6.26), we get

$$\dot{H} + H^2 = \frac{1}{3}\left(\frac{3}{2}\gamma - 1\right)\frac{G(t)\dot{\Lambda}}{\dot{G}} + \frac{1}{3}\Lambda(t) . \tag{6.27}$$

To solve this equation, the derivatives in terms of scale factor a are obtained using the chain rule,

$$\dot{H} = \frac{dH}{dt} = \frac{dH}{da}\frac{da}{dt} = \acute{H}\dot{a} , \tag{6.28}$$

$$\dot{\Lambda} = \frac{d\Lambda}{dt} = \frac{d\Lambda}{da}\frac{da}{dt} = \acute{\Lambda}\dot{a} , \tag{6.29}$$

$$\dot{G} = \frac{dG}{dt} = \frac{dG}{da}\frac{da}{dt} = \acute{G}\dot{a} \tag{6.30}$$

and

$$H = \frac{d}{dt}\ln a = \frac{\dot{a}}{a} \rightarrow \dot{a} = Ha . \tag{6.31}$$

If we substitute the derivatives (6.28), (6.29), (6.30) and (6.31) into equation (6.27), then we obtain

$$\acute{H}\dot{a} + H^{?}a^{?} = \frac{1}{3}\left(\frac{3}{2}\gamma - 1\right)\frac{G(t)\acute{\Lambda}}{\acute{G}} + \frac{1}{3}\Lambda(t) .$$

Multiplying this equation by $\frac{1}{a}$, we get the equation

$$\acute{H}H + \frac{H^2}{a} = \frac{1}{3}\left(\frac{3}{2}\gamma - 1\right)\frac{G(t)\acute{\Lambda}}{\acute{G}a} + \frac{1}{3}\frac{\Lambda(t)}{a} .$$

□

6.2 Solutions of the field equations

In this subsection we discuss the solutions of the field equations for two different early phases: inflationary and radiation dominated.

Equation (6.24), involving H, $\Lambda(t)$, and $G(t)$, admits solutions for the Hubble parameter H if $\Lambda(t)$ and $G(t)$ are given. The possible forms of the gravitational and cosmological constants are discussed in the literature. For example, Dirac suggested that the gravitational constant G varies linearly with the Hubble parameter. [15] built a dimensional argument to justify $\Lambda \sim a^{-2}$. [30] proposed $\Lambda \sim H^2$. Thus, the phenomenological approach to investigating the cosmological constant is generalized to include a term proportional to H^2, time dependent on Λ. [41] obtained the solution of equation (6.24) by taking into account the above assumptions on the gravitational G and cosmological Λ constants.

6.2.1 Dirac's proposition: $G(t) \sim H$

If we accept Dirac's assumption, in which the gravitational constant is linearly proportional to the Hubble parameter, then one may write the following relation between the gravitational constant and the Hubble parameter:

$$G(t) = \alpha H\,. \tag{6.32}$$

Following this, and taking into account the phenomenological approach of [30] to the cosmological constant as $\Lambda \sim H^2$, one can write

$$\Lambda(t) = \beta H^2\,, \tag{6.33}$$

where α and β are dimensionless positive definite constants. Substituting the values $G(t)$ and $\Lambda(t)$ from equations (6.32) and (6.33) into (6.24), we obtain

$$\acute{H} + \left(\frac{\beta+3}{3} - \gamma\beta\right)\frac{H}{a} = 0\,. \tag{6.34}$$

Proof.

$$\acute{H}H + \frac{H^2}{a} = \frac{1}{3}\left(\frac{3}{2}\gamma - 1\right)\frac{G(t)\acute{\Lambda}}{\acute{G}a} + \frac{1}{3}\frac{\Lambda(t)}{a} \tag{6.35}$$

$$\acute{H}H + \frac{H^2}{a} = \frac{1}{3}\left(\frac{3}{2}\gamma - 1\right)\frac{2\beta H^2}{a} + \frac{1}{3}\frac{\beta H^2}{a} \tag{6.36}$$

$$\acute{H} + \frac{H}{a} = \frac{1}{3}\left(\frac{3}{2}\gamma - 1\right)\frac{2\beta H}{a} + \frac{1}{3}\frac{\beta H}{a} \tag{6.37}$$

$$\acute{H} + \frac{H}{a} = \left(\gamma\beta - \frac{2}{3}\beta + \frac{1}{3}\beta\right)\frac{H}{a} \tag{6.38}$$

$$\acute{H} + \left(\frac{\beta+3}{3} - \gamma\beta\right)\frac{H}{a} = 0\,.$$

□

The solution of the first-order differential equation (6.34) is

$$H = \frac{C_0}{a^{\frac{\beta+3}{3}}\left[A(\frac{a}{a_0})^2 + (\frac{a}{a_0})^b\right]^{-\frac{2\beta}{3}}}\,, \tag{6.39}$$

where C_0 is the integration constant.

Proof. To solve the equation, we use the techniques of separation of variables, which leads to

$$\int \frac{dH}{H} = -\int \left(\frac{\beta+3}{3} - \gamma\beta\right)\frac{da}{a}\,. \tag{6.40}$$

Here we should be careful, since the adiabatic parameter is defined as a scale factor dependent function $\gamma(a)$ by equation (6.22). Then substituting (6.22) into equation (6.34), we obtain

$$\int \frac{dH}{H} = -\int \left[\frac{\beta+3}{3} - \beta\left(\frac{4}{3}\frac{A(\frac{a}{a_0})^2 + (\frac{b}{2})(\frac{a}{a_0})^b}{A(\frac{a}{a_0})^2 + (\frac{a}{a_0})^b}\right)\right]\frac{da}{a}\,, \tag{6.41}$$

which leads to immediate evaluation of the left-hand side integral in terms of the Hubble parameter, and the first term of the right-hand side integral in terms of the scale factor,

$$\ln H = -\left(\frac{\beta+3}{3}\right)\ln a + \beta\int\left(\frac{4}{3}\frac{A(\frac{a}{a_0})^2 + (\frac{b}{2})(\frac{a}{a_0})^b}{A(\frac{a}{a_0})^2 + (\frac{a}{a_0})^b}\right)\frac{da}{a}\,. \tag{6.42}$$

To solve the second part of the integral of the right-hand side, we need to define a new parameter K, which is

$$K \equiv \frac{4\beta}{3}\int\left(\frac{A(\frac{a}{a_0})^2 + (\frac{b}{2})(\frac{a}{a_0})^b}{A(\frac{a}{a_0})^2 + (\frac{a}{a_0})^b}\right)\frac{da}{a}\,. \tag{6.43}$$

Then, substituting $z = \frac{a}{a_0}$ and $da = a_0 dz$ in (6.43), the integral K becomes only z dependent,

$$K = \frac{4\beta}{3}\int\left(\frac{Az^2 + (\frac{b}{2})z^b}{Az^2 + z^b}\right)\frac{dz}{z}\,. \tag{6.44}$$

In this integral, by applying polynomial division and then rearranging the parameters, the following from is obtained:

$$K = \frac{4\beta}{3}\int\left(\frac{1}{z} + \left(\frac{b}{2} - 1\right)\frac{z^{b-3}}{A + z^b - 2}\right)dz\,. \tag{6.45}$$

Then we can obtain two integrals

$$K = \frac{4\beta}{3}\int\frac{1}{z}dz + \frac{4\beta}{3}\left(\frac{b}{2} - 1\right)\int\frac{z^{b-3}}{A + z^{b-2}}dz\,. \tag{6.46}$$

Although the first integral is straightforward to evaluate, to solve the second integral on the right-hand side a direct substitution method is used, which is $A + z^{b-2} \equiv u$, $z^{b-3}dz \equiv \frac{du}{b-2}$. Then the integral K gives the following solution:

$$K = \frac{4\beta}{3}\ln z + \beta\frac{2}{3}\ln u + \ln C_0\,. \tag{6.47}$$

The last form of K becomes

$$K = \beta\frac{4}{3}\ln\left|\frac{a}{a_0}\right| + \frac{2\beta}{3}\ln\left|A + \left(\frac{a}{a_0}\right)^{(b-2)}\right| + \ln C_0\,. \tag{6.48}$$

If we insert equation (6.48) into equation (6.42), then equation (6.42) gives the Hubble parameter in the following form:

$$H = \frac{C_0}{a^{\frac{\beta+3}{3}} \left(A(\frac{a}{a_0})^2 + (\frac{a}{a_0})^b\right)^{-\frac{2\beta}{3}}} . \qquad \square$$

If we choose $H = H_0$ and $a = a_0$ in equation (6.39), we have a relation between constants A and C_0, which is given by

$$H_0 = \frac{C_0}{a_0^{\frac{\beta+3}{3}} (A+1)^{-\frac{2\beta}{3}}} . \tag{6.49}$$

The integration constant then becomes

$$C = H_0 a_0^{\frac{\beta+3}{3}} (A+1)^{-\frac{2\beta}{3}} . \tag{6.50}$$

Then by substituting this form of the constant into equation (6.39),

$$H_0 a_0^{\frac{\beta+3}{3}} (A+1)^{-\frac{2\beta}{3}} = \frac{da}{dt} a^{\frac{\beta}{3}} \left(A\left(\frac{a}{a_0}\right)^2 + \left(\frac{a}{a_0}\right)^b\right)^{-\frac{2\beta}{3}} . \tag{6.51}$$

using the property $H = \frac{d}{dt} \ln a = \frac{\dot{a}}{a} = \frac{1}{a}\frac{da}{dt}$, and by integrating the Hubble parameter in terms of time,

$$H_0 a_0^{\frac{\beta+3}{3}} (A+1)^{-\frac{2\beta}{3}} t = \int a^{\frac{\beta}{3}} \left(A\left(\frac{a}{a_0}\right)^2 + \left(\frac{a}{a_0}\right)^b\right)^{-\frac{2\beta}{3}} da , \tag{6.52}$$

we can obtain the scale factor $a(t)$, the Hubble parameter, and the gravitational and cosmological constants, as well as the energy density in both inflationary and radiation dominated regimes. In the following sections we will discuss these regimes in the light of equation (6.52).

6.2.1.1 Inflationary phase

For the inflationary phase $a \ll a_0$, the second term on the right-hand side of the integral in equation (6.52) dominates where $0 < b < 1$ (and in this case $b \neq 0$). These assumptions allow us to reduce equation (6.52) to

$$H_0 a_0^{\frac{\beta+3}{3}} (A+1)^{-\frac{2\beta}{3}} t = \int a^{\frac{\beta}{3}} \left(\frac{a}{a_0}\right)^{-\frac{2\beta}{3} b} da . \tag{6.53}$$

After integrating equation (6.53), the scale factor a can be obtained as

$$a = a_0 \left[\left(\frac{\beta}{3}(1-2b)+1\right) \frac{H_0}{(A+1)^{\frac{2\beta}{3}}} t\right]^{\left(\frac{\beta}{3}(1-2b)+1\right)} . \tag{6.54}$$

The above shows that during inflation, the scale factor of the universe is proportional to

$$a \sim t^{\frac{3}{(1-2b)\beta+3}} , \tag{6.55}$$

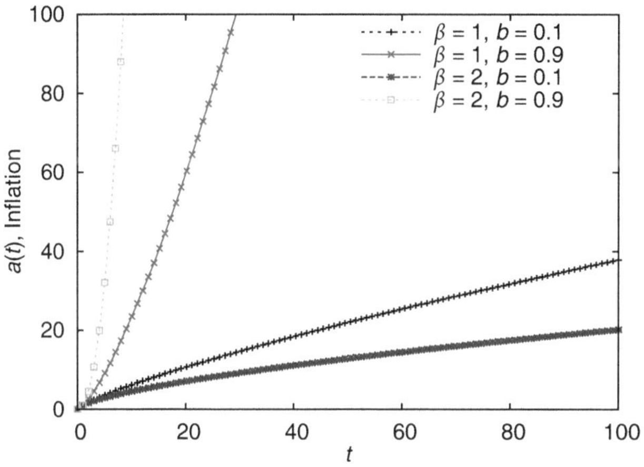

Fig. 6.1: Evolution of the scale factor in the inflationary phase with two different beta parameters, $\beta = 1$ and $\beta = 2$, with two different b parameters, $b = 0.1$ and $b = 0.9$, based on equation (6.55).

which is the power-law inflation [41]. If $\beta = 0$, we can easily see that the scale factor of the universe increases linearly with the age of the universe. Figure 6.1 shows the evolution of the scale factors with different β and b parameters in the inflationary phase based on the linear relation between the gravitational constant and the Hubble parameter ($G(t) \sim H$). As a result, Figure 6.1 shows that in this model, the scale factors with the high b parameter ($b = 0.9$) show exponential expansion. Conversely, the scale factors with a lower b parameter ($b = 0.1$) depict power-law expansion (slow expansion). In addition to this, the scale factors with a higher β value ($\beta = 2$) tend to present steeper exponential expansion compared to the lower beta value ($\beta = 1$) with $b = 0.9$. Conversely, a higher β value ($\beta = 2$) indicates less steep power-law expansion compared to the lower beta value ($\beta = 1$) with $b = 0.1$.

Then, using this scale factor, we can obtain the Hubble parameter

$$H \sim \frac{3}{t\left[(1-2b)\beta + 3\right]} . \tag{6.56}$$

Figure 6.2 shows the evolution of the Hubble parameter of the inflationary model. The figure indicates that the expansion rate of the model becomes higher with higher β values with a fixed b parameter. Similarly, the expansion rates become higher with higher b values for a fixed β parameter.

As a result, the gravitational constant becomes

$$G \sim \frac{3\alpha}{t\left[(1-2b)\beta + 3\right]} . \tag{6.57}$$

As is seen from equation (6.57), the gravitational constant is inversely proportional to the age of the universe. This means that as the age of the universe increases the

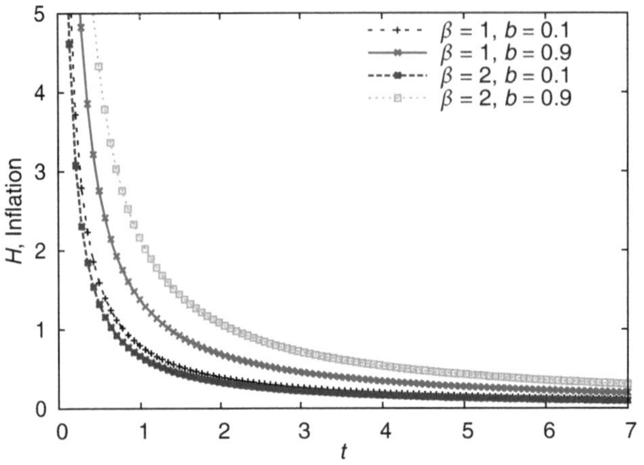

Fig. 6.2: Dynamical behavior of the Hubble parameter in the inflationary phase with two different beta parameters, $\beta = 1$ and $\beta = 2$, with two different b parameters, $b = 0.1$ and $b = 0.9$, based on equation (6.56).

gravitational constant tends to become smaller. Apart from the Hubble parameter and the gravitational constant, the cosmological constant can be found,

$$\Lambda \sim \frac{9\beta}{t^2[(1-2b)\beta+3]^2} \tag{6.58}$$

as well as the energy density,

$$\rho \sim \frac{-3}{4\pi}\frac{\beta}{\alpha}\frac{1}{t\left((1-2b)\,\beta+3\right)}\,. \tag{6.59}$$

As can be seen, the density equation accepts a negative sign. To obtain a model that is consistent with observational evidence, we should choose energy density that is positive definite. This condition puts limits on parameters b and β. These limits are:

- $b \lnsim 0.5$, which forces us to change the interval of the parameter b to $0.5 < b < 1$
- Taking into account this new interval of the parameter b, β becomes $\beta > \frac{-3}{1-2b}$.

As a result, the energy density becomes positive definite.

6.2.1.2 Radiation dominated phase

When the universe enters a radiation dominated phase, the scale factor dominates the initial expansion, $a \gg a_0$. As a result, the first term on the right-hand side of the integral (6.52) becomes dominant:

$$H_0 a_0^{\frac{\beta+3}{3}}(A+1)^{-\frac{2\beta}{3}}t = \int a^{\frac{\beta}{3}}\left(A\left(\frac{a}{a_0}\right)\right)^{-\frac{4\beta}{3}} da\,. \tag{6.60}$$

After taking the integral in terms of the scale factor, we obtain

$$H_0 a_0^{\frac{\beta+3}{3}} (A+1)^{-\frac{2\beta}{3}} t = \frac{a^{-\beta+1}}{1-\beta} A^{\frac{-2\beta}{3}} a_0^{\frac{4\beta}{3}} \tag{6.61}$$

which leads to the scale factor in the radiation dominated phase,

$$a = a_0 \left((1-\beta) \left(\frac{A}{1+A} \right)^{\frac{2\beta}{3}} H_0 t \right)^{\frac{1}{(1-\beta)}} . \tag{6.62}$$

Equation (6.62) shows that the expansion of the universe is proportional to

$$a \sim t^{\frac{1}{(1-\beta)}} . \tag{6.63}$$

Therefore the Hubble parameter is

$$H = \frac{1}{t(1-\beta)} . \tag{6.64}$$

Then the gravitational constant becomes

$$G = \frac{\alpha}{t(1-\beta)} . \tag{6.65}$$

The cosmological constant is obtained as

$$\Lambda = \frac{\beta}{t^2(1-\beta)^2} , \tag{6.66}$$

which leads to the energy density,

$$\rho = \frac{-1}{4\pi} \frac{\beta}{\alpha} \frac{1}{t(1-\beta)} , \tag{6.67}$$

where the parameter β must be greater than one as was stated by the model (6.33). We must however be aware of the fact that Dirac proposed that the gravitational constant and the Hubble parameter might be proportional with a positive constant α, not the constant β. As was stated before, [12] suggested that the cosmological constant may vary with time, and later on, [13] made an assumption that the cosmological constant was proportional to the square of the Hubble parameter with a positive constant β to make the density positive definite. However, $\beta > 0$ in equation (6.33) forces the other dynamical parameters to be negative, such as the gravitational and cosmological constants as well as the Hubble parameter. Making these dynamical parameters negative does not provide a reasonable model compatible with observations. As a result, in the following section we will investigate possible outcomes of another model of the FRW universe with time dependent cosmological constants.

6.2.2 $G(t) \sim 1/H$

In this section, changing Dirac's proposition and assuming that the gravitational constant is inversely proportional to the Hubble parameter by following [41], we will try to obtain an observationally acceptable model. In this framework, the relation between the gravitational constant and the Hubble parameter is given as

$$G = \frac{\alpha}{H}, \quad \alpha > 0 \tag{6.68}$$

and the cosmological constant is again proportional to the square of the Hubble parameter as in the previous model

$$\Lambda(t) = \beta H^2, \quad \beta > 0. \tag{6.69}$$

Substituting equations (6.68) and (6.69) into (6.24), the following equation is obtained:

$$\acute{H} + \frac{H}{a}(1 - \beta + \gamma\beta) = 0. \tag{6.70}$$

Proof. First, using the chain rule, we obtain the first derivatives of the cosmological and gravitational constants in terms of the scale factor as follows:

$$\acute{\Lambda} = \frac{d\Lambda}{da} = \frac{d\Lambda}{dt}\frac{dt}{da} = \frac{d}{dt}\left(\beta H^2\right)\frac{1}{\dot{a}} = 2\beta H\dot{H}\frac{1}{\dot{a}} \tag{6.71}$$

and

$$\acute{G} = \frac{dG}{da} = \frac{dG}{dt}\frac{dt}{da} = \frac{d}{dt}\left(\frac{\alpha_1}{H}\right)\frac{1}{\dot{a}} = -\alpha\frac{\dot{H}}{H^2}\frac{1}{\dot{a}}. \tag{6.72}$$

Substituting these new derivatives (6.71) and (6.72) into equation (6.24), the new form of the equation in terms of the scale factor is obtained:

$$\acute{H}H + \frac{H^2}{a} = (1 - \gamma)\frac{1}{a}\beta H^2. \tag{6.73}$$

After dividing this equation by H,

$$\acute{H} + \frac{H}{a} = (1 - \gamma)\frac{1}{a}\beta H, \tag{6.74}$$

and arranging equation (6.74), we obtain

$$\acute{H} + \frac{H}{a}(1 - \beta + \gamma\beta) = 0.$$ □

The solution of the differential equation (6.70) that governs the dynamical characteristics of the Hubble parameter in terms of the scale factor is

$$H = \frac{C}{a^{(1-\beta)}\left(A\left(\frac{a}{a_0}\right)^2 + \left(\frac{a}{a_0}\right)^b\right)^{\frac{2}{3}\beta}}, \tag{6.75}$$

where C is the integration constant.

Proof. The first-order differential equation (6.70) can be solved using the separation of variables as follows:

$$\frac{dH}{da} = -(1-\beta+\gamma\beta)\frac{H}{a}, \tag{6.76}$$

which leads to

$$\frac{dH}{H} = -\left(1-\beta+\frac{4\beta}{3}\frac{A\left(\frac{a}{a_0}\right)^2+\left(\frac{b}{2}\right)\left(\frac{a}{a_0}\right)^b}{A\left(\frac{a}{a_0}\right)^2+\left(\frac{a}{a_0}\right)^b}\right)\frac{da}{a}. \tag{6.77}$$

By substituting $\frac{a}{a_0} \equiv z \longrightarrow da \equiv a_0 dz$, as in the previous model, and integrating equation (6.77), we obtain

$$\int\frac{dH}{H} = \int(\beta-1)\frac{da}{a} - \frac{4\beta}{3}\int\frac{Az^2+\left(\frac{b}{2}\right)z^b}{Az^2+z^b}\frac{dz}{z}. \tag{6.78}$$

To get the integration of this equation, we use the same procedure as for equation (6.41), which leads to

$$\ln H = (\beta-1)\ln|a| - \frac{4\beta}{3}\ln|z| - \frac{2\beta}{3}\ln\left|A+z^{b-2}\right| + \ln|C|. \tag{6.79}$$

As a result, equation (6.79) is given in terms of the scale factor as in the following form:

$$H = \frac{C}{a^{(1-\beta)}\left(A\left(\frac{a}{a_0}\right)^2+\left(\frac{a}{a_0}\right)^b\right)^{\frac{2}{3}\beta}}. \qquad \square$$

To obtain the integration constant, let us assume $H = H_0$ and $a = a_0$ as initial value problems in equation (6.75). Then the integration constant becomes

$$C = H_0 a_0^{(1-\beta)}(A+1)^{\frac{2}{3}\beta}. \tag{6.80}$$

Substituting the integration constant derived from the assumed initial conditions in the Hubble parameter, the dynamical equation (6.75) becomes

$$H = \frac{\dot{a}}{a} = \frac{H_0 a_0^{(1-\beta)}(A+1)^{\frac{2}{3}\beta}}{a^{(1-\beta)}\left(A\left(\frac{a}{a_0}\right)^2+\left(\frac{a}{a_0}\right)^b\right)^{\frac{2}{3}\beta}}, \tag{6.81}$$

which gives the following form:

$$H_0 a_0^{(1-\beta)}(A+1)^{\frac{2}{3}\beta}dt = a^{(1-\beta)}\left(A\left(\frac{a}{a_0}\right)^2+\left(\frac{a}{a_0}\right)^b\right)^{\frac{2}{3}\beta}\frac{da}{a}. \tag{6.82}$$

By integrating equation (6.82) we obtain

$$H_0 a_0^{(1-\beta)}(A+1)^{\frac{2}{3}\beta}t = \int\frac{1}{a^\beta}\left(A\left(\frac{a}{a_0}\right)^2+\left(\frac{a}{a_0}\right)^b\right)^{\frac{2}{3}\beta}da. \tag{6.83}$$

This equation will allow us to obtain all the dynamical parameters of the cosmological models by providing the solution of the scale factor, which will be obtained for two early phases of the universe.

6.2.2.1 Inflationary phase

In the case of the inflationary phase, the initial scale parameter dominates the late scale factor, $a \ll a_0$. As a result, equation (6.83) is reduced to a simple form,

$$H_0 a_0{}^{(1-\beta)} (A+1)^{\frac{2}{3}\beta} t = \int \frac{1}{a^\beta} \left(\frac{a}{a_0} \right)^{\frac{2}{3}b\beta} da \,. \tag{6.84}$$

By integrating and rearranging this equation, the scale factor is found as

$$a = \left[H_0 a_0{}^{(1-\beta)} (A+1)^{\frac{2}{3}\beta} \left(\beta \left(\frac{2}{3} b - 1 \right) + 1 \right) t \right]^{\frac{3}{2b\beta+3(1-\beta)}} , \tag{6.85}$$

which simply leads to a power law expansion in the inflationary phase

$$a \sim t^{\frac{1}{\beta\left(\frac{2}{3}b-1\right)+1}} , \tag{6.86}$$

where the β parameter has to be chosen as

$$\beta > \frac{1}{1 - \frac{2}{3}b} \tag{6.87}$$

to obtain positive scale factors indicating expansion. Figure 6.3 presents the evolution of the scale factor in the inflation phase with different b. The β parameter is fixed as one (upper panel) as well as with different β values with the b parameter fixed to 0.9 (lower panel). In the upper panel of Figure 6.3, taking into account the condition on the parameter b with fixed $\beta = 1$ (equation (6.87), and $0 < b < 1$, we here use the values $b = 0.1, 0.3, 0.5$ and 0.9. Based on this, the scale parameters that are modeled with small b values tend to increase sharply by the power-law equation (6.86) compared to the ones with large b values. The lower panel in Figure 6.3 shows that scale factors modeled with relatively lower β parameters increase slowly compared to ones with larger β values. Using the scale factor in equation (6.86) for the inflation phase, the Hubble parameter, or the expansion rate of the inflationary phase, is found as

$$H = \frac{1}{t\left(\beta\left(\frac{2}{3}b - 1\right) + 1\right)} \,. \tag{6.88}$$

Figure 6.4 presents the evolution of the Hubble parameter in the inflationary phase in terms of different b values (upper panel) and β values (lower panel). Therefore, upper panel shows that the Hubble parameter decreases rapidly with higher b values for a fixed β value. In addition, the Hubble parameter increases with higher β values for a fixed b value.

Apart from the Hubble parameter, one can obtain the formulae of the gravitational constant,

$$G = \alpha \left(\beta \left(\frac{2}{3} b - 1 \right) + 1 \right) t \tag{6.89}$$

and the cosmological constant,

$$\Lambda = \frac{\beta}{t^2 \left(\beta\left(\frac{2}{3}b - 1\right) + 1\right)^2} \,. \tag{6.90}$$

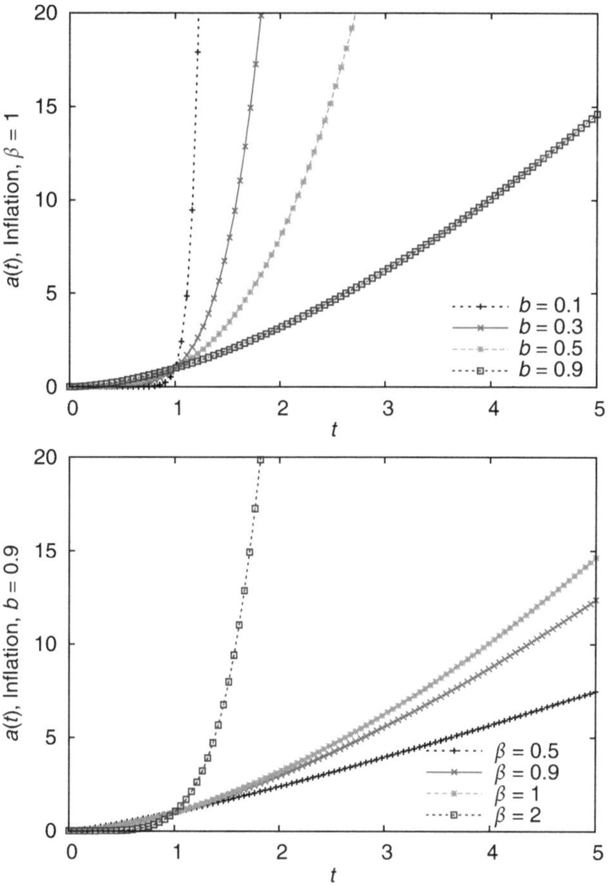

Fig. 6.3: The dynamical behavior of the scale factor in terms of different b values with a fixed β parameter (upper panel), and in terms of different β values with a fixed b parameter (lower panel).

As a result, the energy density in the inflationary phase becomes

$$\rho = \frac{\beta}{4\pi\alpha t^3\left(\beta\left(\frac{2}{3}b - 1\right) + 1\right)^3}, \tag{6.91}$$

which is a positive definite time dependent function.

6.2.2.2 Radiation dominated phase

In the radiation dominated phase of the universe the scale factor dominates the initial scale factor $a \gg a_0$. As a result, the first term on the right-hand side of the integral in equation (6.83) becomes dominant:

$$H_0 a_0{}^{(1-\beta)}(A+1)^{\frac{2}{3}\beta} t = \int \frac{1}{a^\beta} A^{\frac{2}{3}\beta}\left(\frac{a}{a_0}\right)^{\frac{4}{3}\beta} da\,. \tag{6.92}$$

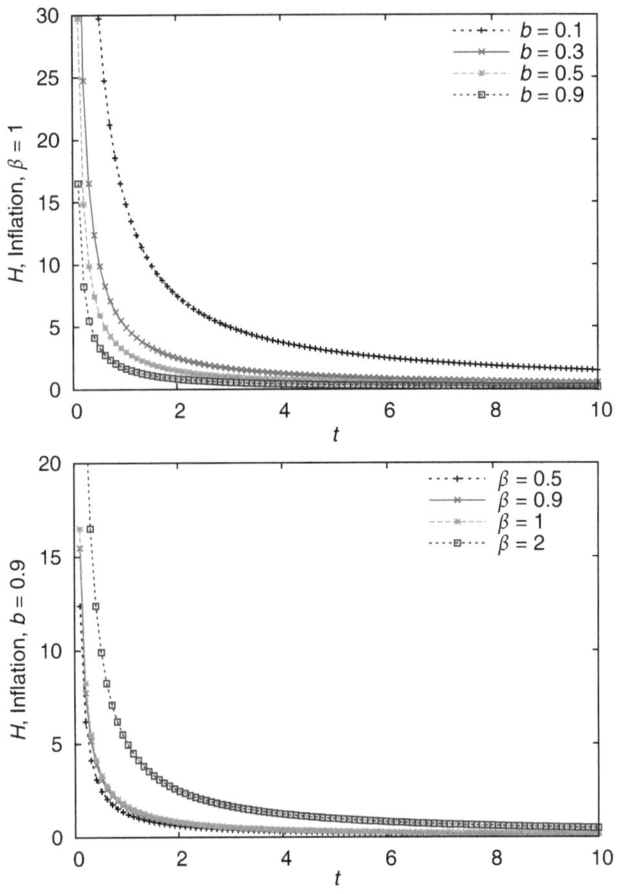

Fig. 6.4: The evolution of the Hubble parameter in terms of different b values with a fixed β parameter (upper panel), and in terms of different β values with a fixed b parameter (lower panel).

After rearranging and integrating in terms of the scale factor a, we obtain the scale factor of the radiation dominated phase for this specific model as follows:

$$a = \left[H_0 a_0^{\frac{3}{\beta+3}} \left(\frac{A+1}{A}\right)^{\frac{2}{3}\beta} \left(\frac{\beta+3}{3}\right) t \right]^{\frac{1}{\frac{\beta}{3}+1}}, \tag{6.93}$$

which shows that the scale factor is proportional to the time with the power-law

$$a \sim t^{\frac{3}{\beta+3}}. \tag{6.94}$$

Figure 6.5 presents the dynamical behavior of the scale factor in the radiation dominated phase in the case of different β parameters. Similar to the inflation phase of this model, the scale factor gradually expands with lower β values compared to higher β

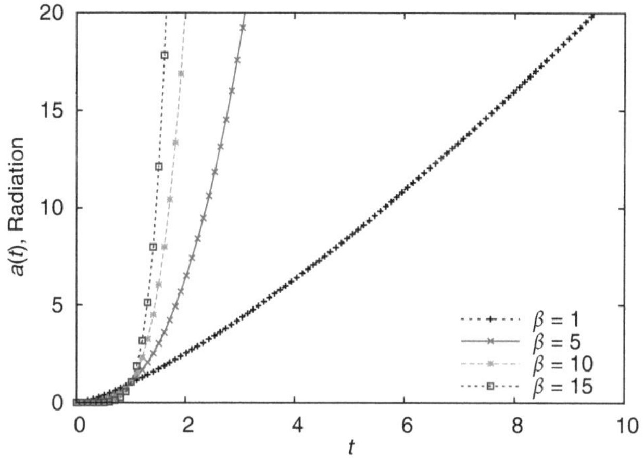

Fig. 6.5: The time evolution of the scale factor in the radiation dominated model with different β parameters.

values. The scale factor (6.93) leads to the equation of the Hubble parameter:

$$H = \frac{1}{t\left(\frac{\beta}{3} + 1\right)} . \tag{6.95}$$

As can be seen, similar to the first model (Dirac's proposition), the Hubble parameter in the radiation dominated phase is independent of the b parameter in the second model. Based on equation (6.95), Figure 6.6 shows how the Hubble parameter evolves in time depending on different β parameters. According to this, again similar

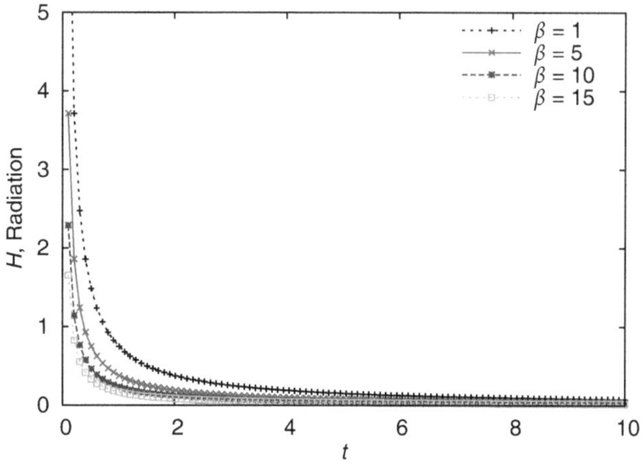

Fig. 6.6: Dynamical evolution of the Hubble parameter in the radiation dominated model with different β parameters.

to the first radiation dominated model, the Hubble parameter decreases and reaches the steady state faster with higher β values.

Using the Hubble parameter, the gravitational constant is obtained:

$$G = \alpha\left(\frac{\beta}{3} + 1\right)t \tag{6.96}$$

as well the cosmological constant

$$\Lambda = \frac{\beta}{t^2\left(\frac{\beta}{3} + 1\right)^2} . \tag{6.97}$$

Finally the energy density becomes

$$\rho = \frac{\beta}{4\alpha\left(\frac{\beta}{3} + 1\right)^3 t^3} , \tag{6.98}$$

which is a positive definite function. In the radiation dominated phase, the gravitational constant increases with cosmic time in both phases, whereas the cosmological constant varies inversely as the square of cosmic time. The energy density varies with the cube of the cosmic time inversely and thus tends to infinity as t approaches the singularity ($t = 0$).

Part II: Variational principle for time dependent oscillations and dissipations

7 Lagrangian and Hamilton descriptions

7.1 Generalized coordinates and velocities

The position of a particle in space at time t is defined by the position vector $\vec{r}(t)$ in Cartesian coordinates, with components $x(t)$, $y(t)$, $z(t)$. The position vector is then written as

$$\vec{r}(t) = x(t)\vec{i} + y(t)\vec{j} + z(t)\vec{k} .$$

The time derivative of the position vector at time t,

$$\vec{v}(t) = \frac{d\vec{r}}{dt} ,$$

is the velocity of the particle, and the second derivative of the position vector is the acceleration of the particle $\vec{a}$,

$$\vec{a}(t) = \frac{d\vec{v}}{dt} = \frac{d^2\vec{r}}{dt^2} .$$

If we consider a N-particle system, rather than a single particle, then it is necessary to specify N position vectors, $\vec{r}_i$, $i = 1, 2, 3, \dots, N$.

Definition 7.1. The number of independent parameters of a system needed to define its configuration is called the *number of degrees of freedom.*

For the system of N particles this number is $3N$. But in general, due to some constraints, the coordinates are not necessarily Cartesian, and this number could be less than $3N$.

Definition 7.2. The k quantities $q_1, q_2, \dots, q_k$, which define the position of a system with k degrees of freedom, are called *generalized coordinates* of the system, and the first derivatives of these quantities in terms of time, $\dot{q}_k$, are called its *generalized velocities.*

7.2 The principle of least action

The most general formulation of the law governing the motion of mechanical systems is the principle of least action, or Hamilton's principle.

Definition 7.3. Every mechanical system is characterized by a definite function $L(q_1, q_2, \dots, q_s \dot{q}_1, \dot{q}_2, \dots, \dot{q}_s, t)$, or, briefly $L(q, \dot{q}, t)$, called the *Lagrangian* of the system. The integral of this function

$$S = \int_{t_1}^{t_2} L\,(q, \dot{q}, t)\, dt , \tag{7.1}$$

https://doi.org/10.1515/9783110515367-007

is called the action. The mechanical system is moving between positions $q^{(1)} = q(t_1)$ and $q^{(2)} = q(t_2)$ in such a way that the action integral takes the least possible value.

From this principle we are going to derive the differential equations, solving the problem of minimization of the integral (7.1). In this derivation we use the system with only one degree of freedom, determined by one function $q(t)$. If $q = q(t)$ is the function that minimizes S, then S would increase when $q(t)$ is replaced by

$$q(t) + \delta q(t) \,, \tag{7.2}$$

where variation of the function $\delta q(t)$ is small everywhere at an interval between t_1 and t_2. Since for $t = t_1$ and $t = t_2$, all functions (7.2) must take the values $q^{(1)}$ and $q^{(2)}$ respectively, it follows that

$$\delta q(t_1) = \delta q(t_2) = 0 \,. \tag{7.3}$$

The change in S,

$$\int_{t_1}^{t_2} L\,(q + \delta q, \dot{q} + \delta\dot{q}, t)\, dt - \int_{t_1}^{t_2} L\,(q, \dot{q}, t)\, dt \,,$$

expanded in powers of δq and $\delta\dot{q}$ in the integrand, in the first order should necessarily vanish, so that the principle of least action becomes

$$\delta S = \delta \int_{t_1}^{t_2} L\,(q, \dot{q}, t)\, dt = 0$$

or

$$\int_{t_1}^{t_2} \left(\frac{\partial L}{\partial q} \delta q + \frac{\partial L}{\partial \dot{q}} \delta \dot{q} \right) dt = 0 \,.$$

Integrating by part

$$\delta S = \left[\frac{\partial L}{\partial \dot{q}} \delta q \right]_{t_1}^{t_2} + \int_{t_1}^{t_2} \left(\frac{\partial L}{\partial q} - \frac{d}{dt} \frac{\partial L}{\partial \dot{q}} \right) \delta q dt = 0 \tag{7.4}$$

and using condition (7.3) we have the integral that must vanish for all values of δq, which implies the Euler–Lagrange equation

$$\frac{d}{dt} \left(\frac{\partial L}{\partial \dot{q}} \right) - \frac{\partial L}{\partial q} = 0 \,.$$

For the system with s degrees of freedom, following the same procedure we then obtain the system of s differential equations

$$\frac{d}{dt} \left(\frac{\partial L}{\partial \dot{q}_i} \right) - \frac{\partial L}{\partial q_i} = 0 \,, \quad i = 1, 2, \ldots s \,. \tag{7.5}$$

For given Lagrangian function these are the ordinary differential equations determining behavior of the mechanical system.

7.3 Hamilton's equations

In the Lagrangian formalism the mechanical state of a system can be described by its generalized coordinates and the velocities. In the previous subsection we give the description of a mechanical system from the Lagrangian perspective. In Hamiltonian formalism, we can represent the same mechanical system in terms of the generalized coordinates and the momenta. We will use the Legendre transformation to formulate the Hamiltonian equation from the Lagrangian of the same system. The Legendre transformation allows us to investigate a dynamical system by changing one set of independent variables to another one. To construct the Hamiltonian, first we obtain the total differential of the Lagrangian, which is function of coordinates, velocities and time

$$dL = \sum_i \frac{\partial L}{\partial q_i} dq_i + \sum_i \frac{\partial L}{\partial \dot{q}_i} d\dot{q}_i + \frac{\partial L}{\partial t} dt \,. \tag{7.6}$$

Introducing the generalized momenta

$$p_i \equiv \frac{\partial L}{\partial \dot{q}_i} \,, \tag{7.7}$$

this expression may be rewritten as

$$dL = \sum_i \left(\frac{\partial L}{\partial q_i} dq_i + p_i d\dot{q}_i \right) + \frac{\partial L}{\partial t} dt \,. \tag{7.8}$$

By using

$$d\left(p_i \dot{q}_i\right) = \dot{q}_i dp_i + p_i d\dot{q}_i \tag{7.9}$$

and substituting $d\left(p_i \dot{q}_i\right) - \dot{q}_i dp_i$ for $p_i d\dot{q}_i$ in the above equation we get

$$dL = \sum_i \left(\frac{\partial L}{\partial q_i} dq_i + d\left(p_i \dot{q}_i\right) - \dot{q}_i dp_i \right) + \frac{\partial L}{\partial t} dt \,. \tag{7.10}$$

From this we get

$$d\left(\sum p_i \dot{q}_i - L\right) = \sum_i \left(\dot{q}_i dp_i - \frac{\partial L}{\partial q_i} \right) - \frac{\partial L}{\partial t} dt \,, \tag{7.11}$$

where the term on the left-hand side determines the Hamiltonian of the system as a function of coordinates and momenta,

$$H\left(p, q, t\right) = \sum_i p_i \dot{q}_i - L \,. \tag{7.12}$$

Substituting

$$\dot{p}_i \equiv \frac{\partial L}{\partial q_i} \,, \tag{7.13}$$

we have the differential form

$$dH = \sum_i (\dot{q}_i dp_i - \dot{p}_i dq_i) - \frac{\partial L}{\partial t} dt\,. \tag{7.14}$$

Differentiating the function $H(p, q, t)$,

$$dH = \sum_i \left(\frac{\partial H}{\partial p_i} dp_i + \frac{\partial H}{\partial q_i} dq_i \right) + \frac{\partial H}{\partial t} dt \tag{7.15}$$

and comparing two differential forms, we obtain Hamilton's equations of motion for variables p and q:

$$\dot{q}_i = \frac{\partial H}{\partial p_i}\,, \quad \dot{p}_i = -\frac{\partial H}{\partial q_i}\,, \quad \frac{\partial H}{\partial t} = -\frac{\partial L}{\partial t}\,. \tag{7.16}$$

These form a set of $2s$ first-order differential equations for the $2s$ unknown functions $p_i(t)$ and $q_i(t)$, which are also called the canonical equations, which replace the s second-order equations in the Lagrangian approach.

If the Hamiltonian does not depend explicitly on time, then

$$\frac{dH}{dt} = \frac{\partial H}{\partial t} = 0 \tag{7.17}$$

and we have the law of conservation of energy.

7.3.1 The Poisson brackets

Definition 7.4. For any two functions of coordinates and momenta, $f(q, p)$ and $g(q, p)$, the Poisson bracket is defined as

$$\{f, g\} = \sum_i \left(\frac{\partial f}{\partial p_i}\frac{\partial g}{\partial q_i} - \frac{\partial f}{\partial q_i}\frac{\partial g}{\partial p_i} \right)\,. \tag{7.18}$$

The Poisson bracket has the following properties:

(a) Antisymmetric

$$\{f, g\} = -\{g, f\}\,.$$

(b) It satisfies the Leibniz identity

$$\{gf_1, f_2\} = g\{f_1, f_2\} + f_1\{g, f_2\}\,.$$

(c) It is bilinear

$$\begin{aligned} \{f_1 + f_2, g\} &= \{f_1, g\} + \{f_2, g\}\,, \\ \{f_1, f_2 + g\} &= \{f_1, f_2\} + \{f_1, g\}\,, \\ \{f_1, cg\} &= c\{f_1, g\}\,, \\ \{cf_1, g\} &= c\{f_1, g\}\,, \end{aligned}$$

where c is a constant.

(d) The Jacobi identity

$$\{f, \{g, h\}\} + \{g, \{h, f\}\} + \{h, \{f, g\}\} = 0\,. \tag{7.19}$$

Taking the partial derivatives of (7.18) in terms of time gives

$$\frac{\partial}{\partial t}\{f, g\} = \left\{\frac{\partial f}{\partial t}, g\right\} + \left\{f, \frac{\partial g}{\partial t}\right\}\,.$$

For a given function of coordinates, momenta and time $f\,(p, q, t)$, the total time derivative is

$$\frac{df}{dt} = \frac{\partial f}{\partial t} + \sum_i \left(\frac{\partial f}{\partial q_i}\dot{q}_i + \frac{\partial f}{\partial p_i}\dot{p}_i\right)\,. \tag{7.20}$$

Inserting $\dot{q}_i$ and $\dot{p}_i$ from Hamilton's equations (7.16) leads to the equation

$$\frac{df}{dt} = \frac{\partial f}{\partial t} + \{H, f\}\,, \tag{7.21}$$

where the Poisson bracket between H and f is

$$\{H, f\} = \sum_i \left(\frac{\partial H}{\partial p_i}\frac{\partial f}{\partial q_i} - \frac{\partial H}{\partial q_i}\frac{\partial f}{\partial p_i}\right)\,. \tag{7.22}$$

Definition 7.5. Functions of the dynamical variables that remain constant during evolution of the system are called *integrals of the motion.*

From (7.21), the condition for the quantity f to be an integral of the motion is

$$\frac{df}{dt} = 0\,, \tag{7.23}$$

or

$$\frac{\partial f}{\partial t} + \{H, f\} = 0\,. \tag{7.24}$$

If the integral of the motion is not explicitly dependent on the time $\partial f/\partial t = 0$, the Poisson bracket between the integral and the Hamiltonian must be zero

$$\{H, f\} = 0\,. \tag{7.25}$$

If one of the functions f and g is either the momenta or coordinates, the Poisson bracket reduces to a partial derivative

$$\{f, p_i\} = -\frac{\partial f}{\partial q_i}\,, \quad \{f, q_i\} = \frac{\partial f}{\partial p_i}$$

and the canonical relations are given by the following equations:

$$\{q_i, q_k\} = 0 = \{p_i, p_k\}\,, \quad \{p_i, q_k\} = \delta_{ik}\,.$$

Theorem 7.1 (Poisson's theorem). *If f and g are two integrals of the motion, their Poisson bracket is an integral of the motion,*

$$\{f, g\} = constant\,.$$

8 Damped oscillator: classical and quantum theory

8.1 Damped oscillator

Definition 8.1. The motion equation of the *damped harmonic oscillator* (DHO) is in the form

$$m\ddot{x} + \gamma\dot{x} + kx = 0\,, \tag{8.1}$$

where m is mass, γ is damping coefficient and k is spring constant. The damped harmonic oscillator equation (8.1) is solved in the form $x = e^{\lambda t}$, where λ satisfies the characteristic equation

$$m\lambda^2 + \gamma\lambda + k = 0. \tag{8.2}$$

The roots of this equation

$$\lambda = \frac{-\gamma \pm \sqrt{\gamma^2 - 4mk}}{2m}\,, \tag{8.3}$$

depending on the parameters, gives three cases:
(a) $\gamma^2 - 4mk > 0\,,$ overdamping case, with two real roots,
(b) $\gamma^2 - 4mk = 0\,,$ critical damping case, with degenerate real roots,
(c) $\gamma^2 - 4mk < 0\,,$ underdamping case, with two complex roots.

The solutions of equation (8.1) for these three cases are
(a) Overdamping:

$$x(t) = e^{-\frac{\gamma}{2m}t}\left(C_1 \sinh\frac{\sqrt{\gamma^2 - 4mk}}{2m}t + C_2 \cosh\frac{\sqrt{\gamma^2 - 4mk}}{2m}t\right) \tag{8.4}$$

(b) Critical damping:

$$x(t) = e^{-\frac{\gamma}{2m}t}\,(C_1 + C_2 t) \tag{8.5}$$

(c) Underdamping:

$$x(t) = e^{-\frac{\gamma}{2m}t}\left(C_1 \sin\frac{\sqrt{4mk - \gamma^2}}{2m}t + C_2 \cos\frac{\sqrt{4mk - \gamma^2}}{2m}t\right). \tag{8.6}$$

8.2 Dissipation in generalized analytical mechanics

The Lagrangian function for the system of coupled linear oscillators with generalized coordinates $q_1, \ldots, q_s$ is a quadratic form in these coordinates and in generalized velocities is $\dot{q}_1, \ldots, \dot{q}_s$. To describe dissipative oscillators, this quadratic form should

https://doi.org/10.1515/9783110515367-008

be considered as a general form with coefficients depending on time:

$$L = \frac{1}{2} \sum_{i,j=1}^{s} \left(m_{ij}(t)\dot{q}_i\dot{q}_j + 2\gamma_{ij}\dot{q}_i q_j - k_{ij}(t) q_i q_j \right) , \tag{8.7}$$

where the mass tensor $m_{ij} = m_{ji}$, and the coupling tensor $k_{ij} = k_{ji}$, are symmetric, but not necessarily positive definite. Tensor γ_{ij} can also be a function of time, but we restrict ourselves here to constant coefficients. Moreover, we can restrict these tensors to being skew-symmetric $\gamma_{ji} = -\gamma_{ij}$, since the symmetric part of the tensor will add and unessential extra total time derivative term to the Lagrangian. Thus we have Lagrangian

$$L = \frac{1}{2} \sum_{i,j=1}^{s} \left(m_{ij}(t)\dot{q}_i\dot{q}_j + \gamma_{ij}(\dot{q}_i q_j - q_i\dot{q}_j) - k_{ij}(t) q_i q_j \right) . \tag{8.8}$$

8.2.1 One degree of freedom

In this case $\gamma_{ij} = 0$ and we choose time dependent mass $m(t) = m_0 f(t)$ and the coupling constant $k(t) = k_0 f(t)$ in its simplest form, depending on only one function $f(t)$. The Lagrangian

$$L = f(t) \left(\frac{m_0}{2}\dot{q}^2 - \frac{m\omega_0^2}{2} q^2 \right) , \tag{8.9}$$

where $\omega_0^2 = k_0/m_0$, gives the equation of motion for the damped harmonic oscillator

$$\ddot{q} + \Gamma(t)\dot{q} + \omega^2 q = 0 . \tag{8.10}$$

Here, time dependent function

$$\Gamma(t) = \frac{\dot{f}}{t} = \frac{d}{dt} \ln m(t)$$

characterizes the dissipation rate.

8.2.1.1 Caldirola–Kanai Lagrangian

In a special case, when Γ is a constant, the corresponding function f depends on time as

$$f(t) = e^{\Gamma t} , \quad \Gamma = \frac{\gamma}{m_0} . \tag{8.11}$$

This corresponds to the so-called Caldirola–Kanai Lagrangian

$$L = e^{\frac{\gamma}{m_0} t} \left(\frac{m_0}{2}\dot{q}^2 - \frac{m_0\omega_0^2}{2} q^2 \right) . \tag{8.12}$$

8.2.2 Two degrees of freedom

We consider the simplest dissipative system with two degrees of freedom, with coordinates q_1, q_2 and velocities $\dot{q}_1, \dot{q}_2$. We suppose that m_{ij} and k_{ij} are time independent. The corresponding Lagrangian is

$$\begin{aligned} L = & \frac{1}{2}\left(m_{11}\dot{q}_1^2 + 2m_{12}\dot{q}_1\dot{q}_2 + m_{22}\dot{q}_2^2\right) + \frac{\gamma}{2}(q_1\dot{q}_2 - \dot{q}_1\dot{q}_2) \\ & - \frac{1}{2}\left(k_{11}q_1^2 + 2k_{12}q_1\dot{q}_2 + k_{22}q_2^2\right) . \end{aligned}$$

In the particular case of vanishing diagonal terms $m_{11} = m_{22} = 0$ and $k_{11} = k_{22} = 0$, denoting $m_{12} \equiv m$ and $k_{12} = k$, we get

$$L = m\dot{q}_1\dot{q}_2 + \frac{\gamma}{2}(q_1\dot{q}_2 - \dot{q}_1q_2) - kq_1q_2 . \tag{8.13}$$

From the Euler–Lagrange equations we then get equations of motion for the damped oscillator and its time reversed image

$$\begin{aligned} m\ddot{q}_1 + \gamma\dot{q}_1 + kq_1 &= 0 , \\ m\ddot{q}_2 - \gamma\dot{q}_2 + kq_2 &= 0 . \end{aligned} \tag{8.14}$$

The two examples above give us different possibilities to describe the damped harmonic oscillator. In the first case we have the system with one degree of freedom, but time dependent parameters. In the second one, we have an additional degree of freedom, leading to a doubling of the degrees of freedom by the time reversed system.

8.3 Bateman Dual Description

A dissipative system is physically incomplete. As a result, an additional equation is needed to derive the defining equations from a variational principle, which is called the Bateman dual description [5]. To set up the canonical formalism for dissipative systems, the doubling of degrees of freedom is required by completing the given dissipative system with its time reversed image. In this way a globally closed system is formed for which the Lagrangian formalism is well defined.

The Bateman dual formalism for obtaining a complementary amplified oscillator from the damped oscillator is illustrated by the following example. Consider for simplicity a single damped oscillator equation

$$m\ddot{x} + \gamma\dot{x} + kx = 0 ,$$

where the coefficients m, γ and k are constants. This equation is derivable from the variational principle for two functions $x(t)$ and $y(t)$,

$$\delta \int y\,(m\ddot{x} + \gamma\dot{x} + kx) = 0\,, \tag{8.15}$$

$$\int \delta y\,(m\ddot{x} + \gamma\dot{x} + kx) + y\delta\left(m\frac{d^2}{dt^2}x + \gamma\frac{d}{dt}x + kx\right) = 0\,, \tag{8.16}$$

$$\int \delta y\,(m\ddot{x} + \gamma\dot{x} + kx) + y\left(m\frac{d^2}{dt^2}\delta x + \gamma\frac{d}{dt}\delta x + k\delta x\right) = 0\,, \tag{8.17}$$

and

$$\int \delta y\,(m\ddot{x} + \gamma\dot{x} + kx) + \left(m\frac{d^2}{dt^2}y - \gamma\frac{d}{dt}y + ky\right)\delta x = 0\,, \tag{8.18}$$

in which both x and y are to be varied. This principle additionally gives the complementary equation

$$m\ddot{y} - \gamma\dot{y} + ky = 0\,. \tag{8.19}$$

Then the dual description is given by the system of equations

$$\begin{aligned} m\ddot{x} + \gamma\dot{x} + kx &= 0\,,\\ m\ddot{y} - \gamma\dot{y} + ky &= 0\,. \end{aligned} \tag{8.20}$$

In this system, the time reversed image of the damped oscillator plays the role of a reservoir or thermal bath into which the energy dissipates from the original system. Thus, the whole system acts as a conservative system [5].

The Lagrangian function for this system,

$$L = m\dot{x}\dot{y} + \frac{\gamma}{2}(x\dot{y} - \dot{x}y) - kxy\,, \tag{8.21}$$

determines the generalized momenta p_x and p_y in terms of x and y, according to:

$$p_x = \frac{\partial L}{\partial \dot{x}} = m\dot{y} - \frac{\gamma}{2}y\,, \quad p_y = \frac{\partial L}{\partial \dot{y}} = m\dot{x} + \frac{\gamma}{2}x\,. \tag{8.22}$$

From the Legendre transformation,

$$H(x, y, p_x, p_y) = \dot{x}p_x + \dot{y}p_y - L\,(x, y, \dot{x}, \dot{y}, t)$$

by substituting in momenta (8.22), we obtain the energy as

$$H(x, y, p_x, p_y) = \left(m\dot{y} - \frac{\gamma}{2}y\right)\dot{x} + \left(m\dot{x} + \frac{\gamma}{2}x\right)\dot{y} - m\dot{x}\dot{y} - \frac{\gamma}{2}\,(x\dot{y} - \dot{x}y) + kxy\,,$$

or simply

$$H(x, y, p_x, p_y) = m\dot{x}\dot{y} + kxy\,. \tag{8.23}$$

To rewrite this in terms of momenta, we express velocities as follows:

$$m\dot{x} = p_y - \frac{\gamma}{2}x \rightarrow \dot{x} = \frac{1}{m}\left(p_y - \frac{\gamma}{2}x\right), \tag{8.24}$$

$$m\dot{y} = p_x + \frac{\gamma}{2}y \rightarrow \dot{y} = \frac{1}{m}\left(p_x + \frac{\gamma}{2}y\right). \tag{8.25}$$

Substituting (8.24) and (8.25) into equation (8.23), finally we obtain the Hamiltonian of the damped oscillator

$$H(x, y, p_x, p_y) = \frac{p_x p_y}{m} + \frac{\gamma}{2m}(yp_y - xp_x) + xy\left(k - \frac{\gamma^2}{4m}\right). \tag{8.26}$$

8.4 Caldirola–Kanai approach to the damped oscillator

The standard equation of motion for the damped harmonic oscillator (8.10) is written in the following form:

$$\ddot{q} + \Gamma(t)\dot{q} + \omega_0^2 q = 0\,, \tag{8.27}$$

where $\Gamma(t) = \frac{d}{dt}\ln m(t)$. In the case where the mass $m(t)$ is given by $m(t) = m_0 e^{\Gamma t}$ and the frequency ω_0^2 is a constant, the Lagrangian of the oscillator is

$$L = \frac{1}{2}e^{\Gamma t}\left(m_0\dot{q}^2 - m_0\omega^2 q^2\right). \tag{8.28}$$

To construct the Hamiltonian of the oscillator, we need a generalized momentum, which is given by

$$p = \frac{\partial L}{\partial \dot{q}} = e^{\Gamma t} m_0 \dot{q} \tag{8.29}$$

and

$$\frac{p^2 e^{-2\Gamma t}}{{m_0}^2} = \dot{q}^2\,. \tag{8.30}$$

Hence, the Hamiltonian is obtained by the Legendre transformation

$$H(q, p) = \dot{q}p - L(q, \dot{q})\,.$$

If we substitute momentum (8.29) into the Legendre transformation, we obtain

$$H = \dot{q}\left(e^{\Gamma t} m_0 \dot{q}\right) - \frac{1}{2}e^{\Gamma t}\left(m_0\dot{q}^2 - m_0\omega_0^2 q^2\right). \tag{8.31}$$

By substituting generalized velocities (8.30) into (8.31), we get the harmonic oscillator with the time dependent mass described by the Hamiltonian

$$H = \frac{p^2}{2m(t)} + \frac{1}{2}m(t)\omega_0^2 q^2\,, \tag{8.32}$$

where the mass $m(t)$ is given by $m(t) = m_0 e^{\Gamma t}$ with $\Gamma = constant$. The Hamiltonian (8.32) is called the Caldirola–Kanai Hamiltonian [11, 26].

8.5 Quantization of the Caldirola–Kanai damped oscillator with constant frequency and constant damping

In this subsection, we introduce a canonical method to quantize the Caldirola–Kanai damped harmonic oscillator with constant damping Γ and frequency ω_0. To construct the Hamiltonian operator of the oscillator, we define momentum and coordinate operators in the Schrödinger representation as

$$q \to q\,, \quad p \to -i\hbar\frac{\partial}{\partial q}\,. \tag{8.33}$$

The classical Caldirola–Kanai Hamiltonian function was given in (8.32), but in this section for simplicity we assume that $m_0 = 1$ and the Hamiltonian is

$$2H = e^{-\Gamma t}p^2 + e^{\Gamma t}\omega_0^2 q^2\,. \tag{8.34}$$

Evolution of the wave function $\Phi(q,t)$ is then described by the time dependent Schrödinger equation as follows:

$$i\hbar\frac{\partial}{\partial t}\Phi(q,t) = H\Phi(q,t)\,, \tag{8.35}$$

where H is defined by equation (8.34). The corresponding Schrödinger equation

$$i\hbar\frac{\partial\Phi}{\partial t} = \left(-\frac{\hbar^2}{2}e^{-\Gamma t}\frac{\partial^2}{\partial q^2} + \frac{1}{2}\omega_0^2 q^2 e^{\Gamma t}\right)\Phi \tag{8.36}$$

is the linear Schrödinger equation with time dependent coefficients. By changing coordinate q and time t in this equation to new coordinate q' and new time t', we can reduce equation (8.36) to the standard Schrödinger equation for the harmonic oscillator. We use the following transformation:

$$q\left(q',t\right) \equiv e^{-\frac{\Gamma}{2}t}q'\,, \quad t \equiv t'\,, \tag{8.37}$$

so that the partial derivatives are then

$$\frac{\partial}{\partial q} = \frac{\partial q'}{\partial q}\frac{\partial}{\partial q'} + \frac{\partial t'}{\partial q}\frac{\partial}{\partial t'} = e^{\frac{\Gamma}{2}t'}\frac{\partial}{\partial q'} \tag{8.38}$$

and

$$\frac{\partial^2}{\partial q^2} = e^{\frac{\Gamma}{2}t'}\frac{\partial}{\partial q'}e^{\frac{\Gamma}{2}t'}\frac{\partial}{\partial q'} = e^{\Gamma t'}\frac{\partial^2}{\partial q'^2} \tag{8.39}$$

and

$$\frac{\partial}{\partial t} = \frac{\partial t'}{\partial t}\frac{\partial}{\partial t'} + \frac{\partial q'}{\partial t}\frac{\partial}{\partial q'} = \frac{\partial}{\partial t'} + \frac{\Gamma}{2}q'\frac{\partial}{\partial q'}\,. \tag{8.40}$$

The Jacobian matrix of the transformation is

$$J = \begin{pmatrix} \frac{\partial q'}{\partial q} & \frac{\partial q'}{\partial t} \\ \frac{\partial t'}{\partial q} & \frac{\partial t'}{\partial t} \end{pmatrix} = \begin{pmatrix} e^{\frac{\Gamma}{2}t} & 0 \\ -\frac{\Gamma}{2}q' & 1 \end{pmatrix} \tag{8.41}$$

and the determinant of the Jacobian matrix becomes

$$J = \begin{vmatrix} e^{\frac{\Gamma}{2}t} & q' \\ 0 & 1 \end{vmatrix} = e^{\frac{\Gamma}{2}t} . \tag{8.42}$$

Since the determinant (8.42) is never zero, our transformation is not singular. After substituting the transformation (8.37) and its partial derivatives (8.38) and (8.40) into (8.36), we obtain

$$i\hbar \left[\frac{\partial}{\partial t} + \frac{\Gamma}{2} q' \frac{\partial}{\partial q'} \right] \Phi = -\frac{\hbar^2}{2} \frac{\partial^2}{\partial q'^2} \Phi + \frac{1}{2} \omega_0^2 q'^2 \Phi \tag{8.43}$$

or

$$i\hbar \frac{\partial}{\partial t} \Phi = -\frac{\hbar^2}{2} \frac{\partial^2}{\partial q'^2} \Phi + \frac{1}{2} \omega_0^2 q'^2 \Phi - i\hbar \frac{\Gamma}{2} q' \frac{\partial}{\partial q'} \Phi . \tag{8.44}$$

Since $t' = t$, we will not make a difference between them. We can remove the first derivative term by introducing the new momentum operator, defined as

$$p' = -i\hbar \frac{\partial}{\partial q'} , \quad p'^2 = -\hbar^2 \frac{\partial^2}{\partial q'^2} . \tag{8.45}$$

Due to (8.38), these new momentum p' and coordinate q' values are related to the old ones by

$$p' = e^{-\frac{\Gamma}{2}t} p , \quad q' = e^{\frac{\Gamma}{2}t} q , \tag{8.46}$$

and satisfy the following commutation relations:

$$\begin{aligned} \left[p', q'\right] &= -i\hbar , \\ \left[p', q\right] &= -i\hbar e^{\frac{-\Gamma}{2}t} , \\ \left[p', p'\right] &= \left[q', q'\right] = 0 . \end{aligned} \tag{8.47}$$

The Hamiltonian function of equation (8.44) can then be represented in terms of p' and q' as

$$H_1 = \frac{1}{2} \left(p'^2 + \Gamma q' p' \right) + \frac{1}{2} \omega_0^2 q'^2 . \tag{8.48}$$

To complete the square in parenthesis we use the next identity for noncommutative operators (8.47),

$$p'^2 + \Gamma q' p' = \left(p' + \frac{\Gamma}{2} q' \right)^2 + i\hbar \frac{\Gamma}{2} - \frac{\Gamma^2}{4} q'^2 . \tag{8.49}$$

As a result we get a new Hamiltonian H_1 in the form

$$H_1 = \frac{1}{2} \left(p' + \frac{\Gamma}{2} q' \right)^2 + \frac{1}{2} \left(\omega_0^2 - \frac{\Gamma^2}{4} \right) q'^2 + i\hbar \frac{\Gamma}{4} , \tag{8.50}$$

which implies that the new frequency Ω^2 is defined as

$$\Omega^2 \equiv \omega^2 - \frac{\Gamma^2}{4} \,. \tag{8.51}$$

Below we consider only the underdamped case, when $\Gamma < 2\omega_0$ and, as follows, $\Omega^2 > 0$. Hence, equation (8.50) becomes

$$H_1 = \frac{1}{2}\left(p' - \frac{\Gamma}{2}q'\right)^2 + \frac{1}{2}\Omega^2 q'^2 + i\hbar\frac{\Gamma}{4} \tag{8.52}$$

and we can rewrite the time dependent Schrödinger equation as

$$i\hbar\frac{\partial}{\partial t}\Phi = H_1\Phi \,,$$

or in the explicit form

$$i\hbar\left(\frac{\partial}{\partial t} - \frac{\Gamma}{4}\right)\Phi = \frac{1}{2}\left(-i\hbar\frac{\partial}{\partial q'} + \frac{\Gamma}{2}q'\right)^2\Phi + \frac{1}{2}\Omega^2 q'^2\Phi \,. \tag{8.53}$$

By the local gage transformation of the wave function's phase, we can rewrite the first term on the right-hand side of equation (8.53). We have

$$\left(-i\hbar\frac{\partial}{\partial q'} + \frac{\Gamma}{2}q'\right)\Phi = -i\hbar\left(\frac{\partial}{\partial q'} + \frac{i\Gamma}{2\hbar}q'\right)\Phi = -i\hbar e^{-\frac{i\Gamma}{4\hbar}q'^2}\frac{\partial}{\partial q'}\left(\Phi e^{\frac{i\Gamma}{4\hbar}q'^2}\right) \tag{8.54}$$

and the square of (8.54)

$$\left(-i\hbar\frac{\partial}{\partial q'} + \frac{\Gamma}{2}q'\right)^2\Phi = -\hbar^2\left(\frac{\partial}{\partial q'} + \frac{i\Gamma}{2\hbar}q'\right)^2\Phi = -\hbar^2 e^{-\frac{i\Gamma}{4\hbar}q'^2}\frac{\partial^2}{\partial q'^2}\underbrace{\Phi e^{\frac{i\Gamma}{4\hbar}q'^2}}_{K} \,. \tag{8.55}$$

Let us define a new function K, which is a function of q' and t with respect to the right-hand side of equation (8.55),

$$K\left(q', t\right) \equiv \Phi\left(q', t\right) e^{\frac{i\Gamma}{4\hbar}q'^2} \,, \tag{8.56}$$

then for the wave function we have substitution

$$\Phi\left(q', t\right) = K\left(q', t\right)\left(e^{-\frac{i\Gamma}{4\hbar}q'^2}\right) \,. \tag{8.57}$$

The Schrödinger equation for $K\left(q', t\right)$ becomes

$$i\hbar\left(\frac{\partial}{\partial t} - \frac{\Gamma}{4}\right)K = -\frac{1}{2}\hbar^2\frac{\partial^2}{\partial q'^2}K + \frac{1}{2}\Omega^2 q'^2 K \,. \tag{8.58}$$

This equation can be reduced to the standard harmonic oscillator form. For separation of the time and the space variables we use substitution

$$K\left(q', t\right) = e^{-\frac{i}{\hbar}\epsilon t}\xi\left(q'\right) \tag{8.59}$$

and for time independent wave function $\xi(q')$ we find ordinary the differential equation

$$\left(\epsilon - \frac{i\hbar}{4}\Gamma\right)\xi = -\frac{\hbar^2}{2}\xi'' + \frac{1}{2}\Omega^2 {q'}^2 \xi \,. \tag{8.60}$$

We multiply this equation with $-\frac{2}{\hbar^2}$,

$$-\frac{2}{\hbar^2}\left(\epsilon - \frac{i\hbar}{4}\Gamma\right)\xi = \xi'' - \frac{1}{\hbar^2}\Omega^2 {q'}^2 \xi \,, \tag{8.61}$$

and rescale q' as follows:

$$q' \equiv ay \to \frac{\partial}{\partial {q'}^2} = \frac{1}{a^2}\frac{\partial}{\partial y^2} \,. \tag{8.62}$$

Equation (8.61) then becomes

$$-\frac{2}{\hbar^2}a^2\left(\epsilon - \frac{i\hbar}{4}\Gamma\right)\xi = \frac{\partial^2}{\partial y^2}\xi - \frac{1}{\hbar^2}\Omega^2 a^4 y^2 \xi \,. \tag{8.63}$$

We denote

$$a \equiv \sqrt{\frac{\hbar}{\Omega}} \tag{8.64}$$

so that equation (8.63) is reduced to the following form:

$$-\frac{2}{\hbar\Omega}\left(\epsilon - \frac{i\hbar}{4}\Gamma\right)\xi = \frac{\partial^2}{\partial y^2}\xi - y^2\xi \,. \tag{8.65}$$

If we introduce new parameter $s \equiv \frac{2}{\hbar\Omega}\left(\epsilon - \frac{i\hbar}{4}\Gamma\right)$, the stationary Schrödinger equation (8.65) becomes

$$\xi'' + \left(s - y^2\right)\xi = 0 \,. \tag{8.66}$$

This differential equation can be transformed to the Hermite differential equation. By making the substitution

$$\xi(y) = e^{-\frac{y^2}{2}} f(y) \tag{8.67}$$

and taking the first and the second derivatives with respect to y we have

$$\begin{aligned} \xi'(y) &= \left[f' - yf\right]e^{-\frac{y^2}{2}} \,, \\ \xi''(y) &= \left[f'' - 2yf' - f + y^2 f\right]e^{-\frac{y^2}{2}} \,. \end{aligned} \tag{8.68}$$

Substitution of the first and second derivatives into (8.66) gives the Hermite differential equation

$$f'' - 2yf' + (s-1)f = 0 \,. \tag{8.69}$$

If $s - 1 = 2n$, where $n = 1, 2, \dots$ are positive integer numbers, equation (8.69) admits solutions in the form of the Hermite polynomials. In this case we have the discrete set of the energy eigenvalues

$$s = 2n + 1 \equiv \frac{2}{\hbar\Omega}\left(\epsilon - \frac{i\hbar}{4}\Gamma\right) \to \tag{8.70}$$

in which ϵ_n is

$$\epsilon_n = \hbar\Omega\left(n + \frac{1}{2}\right) + \frac{i\hbar}{4}\Gamma\,. \tag{8.71}$$

We can see that the energy eigenvalues are complex numbers $\epsilon_n = E_n + \frac{i\hbar}{4}\Gamma$, with the real part as the harmonic oscillator spectrum,

$$E_n = \hbar\Omega\left(n + \frac{1}{2}\right)\,, \tag{8.72}$$

determined by effective frequency Ω. The imaginary part of the spectrum depends on damping parameter Γ and implies that corresponding eigenfunctions would grow with time.

The Hermite differential equation is solved by applying a series solution as

$$\begin{aligned} f(y) &= \sum_{n=0}^{\infty} c_n y^n\,,\\ f'(y) &= \sum_{n=1}^{\infty} c_n n y^{n-1}\,,\\ f''(y) &= \sum_{n=2}^{\infty} c_n n(n-1) y^{n-2}\,. \end{aligned} \tag{8.73}$$

If we substitute them into the Hermite differential equation

$$\sum_{n=2}^{\infty} c_n n(n-1)y^{n-2} - 2y\sum_{n=1}^{\infty} c_n n y^{n-1} + (s-1)\sum_{n=0}^{\infty} c_n y^n = 0 \tag{8.74}$$

and denote $s - 1 = \lambda$, then

$$\sum_{n=0}^{\infty} c_{n+2} n(n+2)(n+1)y^n - 2\sum_{n=1}^{\infty} c_n n y^n + \lambda\sum_{n=0}^{\infty} c_n y^n = 0 \tag{8.75}$$

and

$$(2c_2 + \lambda c_0) + \sum_{n=1}^{\infty}\left[(n+1)(n+2)c_{n+2} - 2nc_n + \lambda c_n\right] y^n = 0\,. \tag{8.76}$$

By using the recurrence relation, we get

$$c_{n+2} = \frac{2n - \lambda}{(n+1)(n+2)} c_n \tag{8.77}$$

and

$$2c_2 + \lambda c_0 = 0\,, \quad \rightarrow \quad c_2 = -\frac{\lambda}{2} c_0\,. \tag{8.78}$$

This relation is just a special case of the first general recurrence relation for $n = 0, 1, \ldots$. To obtain the linearly independent solutions, we first choose $c_0 = 0$, $c_1 = 1$. Hence, the surviving terms are

$$\begin{aligned} c_3 &= \frac{2-\lambda}{3!} c_1\\ c_5 &= \frac{(6-\lambda)(2-\lambda)}{5!} c_1 \ldots \end{aligned} \tag{8.79}$$

and so on, but the even integers are zero because of $c_0 = 0$. Thus, the first linearly independent solution is

$$f_1 = c_1 \left[y + \frac{2-\lambda}{3!} y^3 + \frac{(6-\lambda)(2-\lambda)}{5!} y^5 + \dots \right] . \tag{8.80}$$

The surviving coefficients for the second linearly independent solution with respect to $c_1 = 0$, $c_0 = 1$ are

$$\begin{aligned} c_2 &= -\frac{\lambda}{2!} c_0 \\ c_4 &= -\frac{(4-\lambda)\lambda}{4!} c_0 \\ c_6 &= -\frac{(8-\lambda)(4-\lambda)\lambda}{6!} c_0 \dots \end{aligned} \tag{8.81}$$

Hence,

$$f_2 = c_0 \left[1 - \frac{\lambda}{2!} y^2 - \frac{(4-\lambda)\lambda}{4!} y^4 - \frac{(8-\lambda)(4-\lambda)\lambda}{6!} y^6 \dots \right] . \tag{8.82}$$

Using the superposition principle, these two solutions can be combined as the general solution of the Hermite equation

$$\begin{aligned} f &= c_0 \left[1 - \frac{\lambda}{2!} y^2 - \frac{(4-\lambda)\lambda}{4!} y^4 - \frac{(8-\lambda)(4-\lambda)\lambda}{6!} y^6 \dots \right] \\ &+ c_1 \left[y + \frac{2-\lambda}{3!} y^3 + \frac{(6-\lambda)(2-\lambda)}{5!} y^5 + \dots \right] . \end{aligned} \tag{8.83}$$

We can rewrite this solution in terms of the hypergeometric function as follows:

$$f = c_0 \, {}_1F_1 \left(-\frac{1}{4}\lambda; \frac{1}{2}; y^2 \right) + c_1 y \, {}_1F_1 \left(-\frac{1}{4}(\lambda - 2); \frac{3}{2}; y^2 \right) . \tag{8.84}$$

Then, the general solution of the original differential equation is

$$\xi(y) = e^{-\frac{y^2}{2}} \left[c_0 \, {}_1F_1 \left(-\frac{1}{4}\lambda; \frac{1}{2}; y^2 \right) + c_1 H_{\frac{\lambda}{2}}(y) \right] , \tag{8.85}$$

where c_0 and c_1 are constants, $H_{\frac{\lambda}{2}}$ is the Hermite polynomial and ${}_1F_1$ is a confluent hypergeometric function. However, we are only interested in the solution for which $\xi(y) \to 0$ as $y \to \infty$. We have

$$\xi(y) = c_1 e^{-\frac{y^2}{2}} H_{\frac{\lambda}{2}}(y) , \tag{8.86}$$

where $\lambda = 2n$. As a result, the wave function of the n-th energy level becomes

$$\Phi_n(q', t) = c_n e^{\frac{-i}{\hbar}\epsilon_n t} e^{-\frac{i\Gamma}{4\hbar} q'^2} e^{-\frac{y^2}{2}} H_n(y) . \tag{8.87}$$

Coefficient c_n can be fixed by the normalization condition for the wave function

$$\int_{-\infty}^{\infty} |\Phi_n(q, t)|^2 dq = 1 , \tag{8.88}$$

implying

$$c_n{}^2 a \int_{-\infty}^{\infty} H_n^2(y) e^{-y^2} dy = c_n{}^2\, a\, 2^n n! \sqrt{\pi} = 1 \tag{8.89}$$

due to normalization of the Hermite polynomials. As a result

$$c_n = \frac{1}{\sqrt{2^n n! \pi^{1/2} a}}, \tag{8.90}$$

in which

$$a = \sqrt{\frac{\hbar}{\Omega}}.$$

This gives normalized wave function

$$\Phi_n(q,t) = \frac{1}{\sqrt{2^n n! \pi^{1/2} a}} e^{\frac{-i}{\hbar}\epsilon_n t} e^{-\frac{i\Gamma}{4\hbar} q^2 e^{\Gamma t}} e^{-\frac{q^2}{2a^2} e^{\Gamma t}} H_n\left(\frac{q}{a} e^{\frac{\Gamma}{2}t}\right), \tag{8.91}$$

where

$$\epsilon_n = \hbar\Omega\left(n + \frac{1}{2}\right) + \frac{i\hbar}{4}\Gamma.$$

It is here convenient to introduce the time dependent function

$$\sigma(t) \equiv a e^{-\frac{\Gamma}{2}t} = \sqrt{\frac{\hbar}{\Omega}} e^{-\frac{\Gamma}{2}t} \tag{8.92}$$

and rewrite our wave function in the form

$$\Phi_n(q,t) = \frac{1}{\sqrt{2^n n! \pi^{1/2} \sigma(t)}} e^{\frac{-i}{\hbar}E_n t} e^{-\frac{i\Gamma}{4\Omega}\frac{q^2}{\sigma^2(t)}} e^{-\frac{1}{2}\frac{q^2}{\sigma^2(t)}} H_n\left(\frac{q}{\sigma(t)}\right), \tag{8.93}$$

where

$$E_n = \hbar\Omega\left(n + \frac{1}{2}\right)$$

is the spectrum of the usual harmonic oscillator. Thus, the function $\sigma(t)$ absorbs all time dependence related with damping.

9 Sturm–Liouville problem as a damped oscillator with time dependent damping and frequency

9.1 Sturm–Liouville problem in double oscillator representation and self-adjoint form

The general form of the linear operator that corresponds to the second-order differential equations is given by

$$\mathcal{L} = p_0(t)\frac{d^2}{dt^2} + p_1(t)\frac{d}{dt} + p_2(t) . \tag{9.1}$$

Here we assume that p_0, p_1 and p_2 are real functions of time t in interval $a \le t \le b$, and the first $2-i$ derivatives of p_i are continuous. Further, if function $p_0(t)$ has singular points, then the interval is chosen as $[a, b]$ so that there are no singular points in the interior of the interval.

Definition 9.1. For a linear operator $\mathcal{L}$ the bilinear form is given by the associated action integral,

$$\begin{aligned} S \equiv \langle v(t)|\mathcal{L}u(t)\rangle &= \int_b^a v(t)\mathcal{L}u(t)dt \\ &= \int_b^a v\,[p_0\ddot{u} + p_1\dot{u} + p_2 u]\,dt , \end{aligned} \tag{9.2}$$

where the dots on the real function $u(t)$ denote derivatives.

If we apply the integration by parts to the associated action integral (9.2), then

$$\begin{aligned} S = \langle \mathcal{L}v(t)|u(t)\rangle &= \left[vp_1 u + v\dot{p}_0\dot{u} - \frac{d}{dt}(vp_0)\,u \right]_{t=a}^{b} \\ &+ \int_b^a \left[\frac{d^2}{dt^2}(vp_0)\,u - \frac{d}{dt}(vp_1)\,u + p_2 vu \right] dt , \end{aligned} \tag{9.3}$$

which leads to

$$\begin{aligned} S = \langle \mathcal{L}v(t)|u(t)\rangle &= \left[vp_1 u + v\dot{p}_0\dot{u} - \frac{d}{dt}(vp_0)\,u \right]_{t=a}^{b} \\ &+ \int_b^a u\,[\ddot{v}p_0 + \dot{v}\,(2\dot{p}_0 - p_1) + (p_2 - \dot{p}_1 + \ddot{p}_0)\,v]\,dt . \end{aligned} \tag{9.4}$$

Here the nonintegral part vanishes due to the boundary values and the self-adjoint reduction.

https://doi.org/10.1515/9783110515367-009

If $p_1 = \dot{p}_0$ in equation (9.4), we can easily see that

$$\langle v(t)|\mathcal{L}u(t)\rangle = \langle \mathcal{L}v(t)|u(t)\rangle = S\,. \tag{9.5}$$

Hence, under the condition $p_1 = \dot{p}_0$, equations (9.1) and (9.4) can be written in the following form:

$$\begin{aligned}
\frac{d}{dt}[\dot{u}p_0] + p_2u = 0 \to \mathcal{L} = \frac{d}{dt}\left[p_0\frac{d}{dt}\right] + p_2\,,\\
\frac{d}{dt}[\dot{v}p_0] + p_2v = 0 \to \overline{\mathcal{L}} = \frac{d}{dt}\left[p_0\frac{d}{dt}\right] + p_2\,,
\end{aligned} \tag{9.6}$$

where $\overline{\mathcal{L}}$ is the adjoint operator. In this case operator $\mathcal{L}$ is the self-adjoint operator, $\overline{\mathcal{L}} = \mathcal{L}$, and for action we have $\langle v(t)|\mathcal{L}u(t)\rangle = \langle \mathcal{L}v(t)|u(t)\rangle$. The canonical notation for this operator is

$$\mathcal{L}u = \overline{\mathcal{L}}u = \frac{d}{dt}\left[p(t)\frac{d}{dt}u(t)\right] + q(t)u(t)\,, \tag{9.7}$$

where p_0 is replaced by p and p_2 is replaced by q. If p_1 is not equal to $\dot{p}_0$, $p_1 \neq \dot{p}_0$, from (9.4), we have two different equations,

$$\ddot{u}p_0 + \dot{u}p_1 + p_2u = 0 \tag{9.8}$$

and

$$\ddot{v}p_0 + (2\dot{p}_0 - p_1)\,\dot{v} + (p_2 - \dot{p}_1 + \ddot{p}_0)\,v = 0\,. \tag{9.9}$$

The adjoint operators for these equations are

$$\mathcal{L} = p_0\frac{d^2}{dt^2} + p_1\frac{d}{dt} + p_2 \tag{9.10}$$

and

$$\overline{\mathcal{L}} = p_0\frac{d^2}{dt^2} + (2\dot{p}_0 - p_1)\,\frac{d}{dt} + (p_2 - \dot{p}_1 + \ddot{p}_0.)\,. \tag{9.11}$$

This shows that the operator $\mathcal{L}$ is not self-adjoint. If we divide equations (9.8) and (9.9) by p_0, we obtain

$$\ddot{u} + \frac{p_1}{p_0}\dot{u} + \frac{p_2}{p_0}u = 0 \tag{9.12}$$

and

$$\ddot{v} + \left(\frac{2\dot{p}_0 - p_1}{p_0}\right)\dot{v} + \left(\frac{p_2 - \dot{p}_1 + \ddot{p}_0}{p_0}\right)v = 0\,. \tag{9.13}$$

In terms of new variables $\Gamma \equiv \frac{p_1}{p_0}$ and $\omega^2 \equiv \frac{p_2}{p_0}$, the above equations (9.12) and (9.13) become

$$\ddot{u} + \Gamma(t)\dot{u} + \omega^2(t)u = 0\,, \tag{9.14}$$

$$\ddot{v} + \left(-\Gamma(t) + \frac{2\dot{p}_0}{p_0}\right)\dot{v} + \left(\omega^2(t) - \dot{\Gamma} - \Gamma\frac{\dot{p}_0}{p_0} + \frac{\ddot{p}_0}{p_0}\right)v = 0\,. \tag{9.15}$$

Denoting $\ln p_0 \equiv \Phi(t)$ and substituting this into equation (9.15), we get

$$\ddot{v} + \left(-\Gamma(t) + 2\dot{\Phi}(t)\right)\dot{v} + \left(\omega^2(t) - \dot{\Gamma} - \Gamma\dot{\Phi}(t) + \ddot{\Phi}(t) + \dot{\Phi}^2(t)\right)v = 0\,. \tag{9.16}$$

In the special case $p_0 = 1$, and, as follows, $\Phi = 0$, equations (9.14) and (9.16) are reduced to the following forms:

$$\ddot{u} + \Gamma(t)\dot{u} + \omega^2(t)u = 0\,, \tag{9.17}$$

$$\ddot{v} - \Gamma(t)\dot{v} + \left(\omega^2(t) - \dot{\Gamma}\right)v = 0\,. \tag{9.18}$$

The Lagrangian function for these equations is

$$L = \dot{u}\dot{v} - \frac{1}{2}\Gamma(t)\left(v\dot{u} - \dot{v}u\right) - \left(\omega^2(t) - \frac{1}{2}\dot{\Gamma}(t)\right)uv\,. \tag{9.19}$$

Then the canonical momenta determined by this Lagrangian are

$$p_u = \frac{\partial L}{\partial \dot{u}} = \dot{v} - \frac{1}{2}\Gamma(t)v\,, \quad p_v = \frac{\partial L}{\partial \dot{v}} = \dot{u} + \frac{1}{2}\Gamma(t)u\,. \tag{9.20}$$

Using the Legendre transformation,

$$H = \dot{u}p_u + \dot{v}p_v - L\,,$$

we find the Hamiltonian of the system as

$$H(u, v, p_u, p_v) = p_u p_v + \frac{1}{2}\Gamma\left(vp_v - up_u\right) + \left(\omega^2(t) - \frac{1}{4}\Gamma^2(t) - \frac{1}{2}\dot{\Gamma}(t)\right)uv\,. \tag{9.21}$$

9.1.1 Particular cases for the nonself-adjoint equation

Case I: The damping Γ is a constant and the frequency $\omega^2(t)$ is the generic function of time. We then get the Bateman dual description for the damped harmonic oscillator with time dependent frequency

$$\ddot{u} + \Gamma\dot{u} + \omega^2(t)u = 0$$

and

$$\ddot{v} - \Gamma\dot{v} + \omega^2(t)v = 0\,.$$

The Lagrangian function for this double oscillator system is

$$L = \dot{u}\dot{v} - \frac{1}{2}\Gamma\left(v\dot{u} - \dot{v}u\right) - \omega^2(t)uv\,. \tag{9.22}$$

Therefore, the corresponding momenta are

$$p_u = \frac{\partial L}{\partial \dot{u}} = \dot{v} - \frac{1}{2}\Gamma v\,, \quad p_v = \frac{\partial L}{\partial \dot{v}} = \dot{u} + \frac{1}{2}\Gamma u\,. \tag{9.23}$$

These substituted into

$$H = \dot{u}p_u + \dot{v}p_v - L = \dot{u}\dot{v} + \omega^2(t)uv \tag{9.24}$$

give the Hamiltonian function

$$H(u, v, p_u, p_v) = p_u p_v + \frac{1}{2}\Gamma\,(vp_v - up_u) + \left(\omega^2(t) - \frac{1}{4}\Gamma^2\right) uv\,. \tag{9.25}$$

Case II: The frequency ω^2 is a constant and the damping $\Gamma(t)$ is the generic function of time. The damped oscillator equations then become

$$\ddot{u} + \Gamma(t)\dot{u} + \omega^2 u = 0\,,$$
$$\ddot{v} - \Gamma(t)\dot{v} + \left(\omega^2 - \dot{\Gamma}\right) v = 0\,.$$

The Lagrangian for this time dependent damped oscillator with constant frequency is

$$L = \dot{u}\dot{v} - \frac{1}{2}\Gamma(t)\,(v\dot{u} - \dot{v}u) - \left(\omega^2 - \frac{1}{2}\dot{\Gamma}(t)\right) uv\,. \tag{9.26}$$

The momenta

$$p_u = \frac{\partial L}{\partial \dot{u}} = \dot{v} - \frac{1}{2}\Gamma(t)v\,, \quad p_v = \frac{\partial L}{\partial \dot{v}} = \dot{u} + \frac{1}{2}\Gamma(t)u \tag{9.27}$$

substituted into

$$H = \dot{u}p_u + \dot{v}p_v - L = \dot{u}\dot{v} + \left(\omega^2 - \frac{1}{2}\dot{\Gamma}(t)\right) uv \tag{9.28}$$

give the Hamiltonian function

$$H(u, v, p_u, p_v) = p_u p_v + \frac{1}{2}\Gamma(t)\,(vp_v - up_u) + \left(\omega^2 - \frac{1}{4}\Gamma^2(t) - \frac{1}{2}\dot{\Gamma}(t)\right) uv\,.$$

Case III: The damping Γ and the frequency ω^2 are constants. This is the standard Bateman doubled damped harmonic oscillator with constant damping and constant frequency

$$\ddot{u} + \Gamma\dot{u} + \omega^2 u = 0\,,$$
$$\ddot{v} - \Gamma\dot{v} + \omega^2 v = 0\,.$$

The Lagrangian

$$L = \dot{u}\dot{v} - \frac{1}{2}\Gamma\,(v\dot{u} - \dot{v}u) - \omega^2 uv\,, \tag{9.29}$$

generalized momenta

$$p_u = \frac{\partial L}{\partial \dot{u}} = \dot{v} - \frac{1}{2}\Gamma v\,, \quad p_v = \frac{\partial L}{\partial \dot{v}} = \dot{u} + \frac{1}{2}\Gamma u\,, \tag{9.30}$$

and the Legendre transformation

$$H = \dot{u}p_u + \dot{v}p_v - L = \dot{u}\dot{v} + \omega^2 uv \tag{9.31}$$

give the Hamiltonian

$$H(u, v, p_u, p_v) = p_u p_v + \frac{1}{2}\Gamma\,(vp_v - up_u) + \left(\omega^2 - \frac{1}{4}\Gamma^2\right) uv\,. \tag{9.32}$$

In summary, if a system like a damped parametric oscillator is described by the general second-order differential equation that is not self-adjoint, the variation description of the system includes the adjoint equation, which can be termed a mirror image. The particular case in which the damping term Γ and the frequency ω^2 are constants is the time reversal image of the system, and the energy for the considered system is a conserved quantity.

9.1.2 Variational principle for self-adjoint operator

In the previous subsection, we formulated the variational action functional as a bilinear form for operator $\mathcal{L}$, which describes the doublet oscillator system. If we wish to avoid doubling the degrees of freedom, we should find another Lagrangian formulation that includes variables of original system only. To do this, we make note the well-known fact that by using the integration factor, any linear second-order equation can be written in the self-adjoint form, which is called the Sturm–Liouville form. If we consider the generic parametric damped oscillator, represented as

$$\ddot{u} + \Gamma(t)\dot{u} + \omega^2(t)u = 0\,,$$

then the corresponding linear operator

$$\mathcal{L} = \frac{d^2}{dt^2} + \Gamma(t)\frac{d}{dt} + \omega^2(t) \tag{9.33}$$

is not self-adjoint. To obtain the self-adjoint operator, we need to find an integration factor. If we multiply equation (9.14) with the integration factor $\mu(t)$, so that

$$\mu\left[\frac{d^2}{dt^2}u + \Gamma(t)\frac{d}{dt}u + \omega^2(t)u\right] = 0\,, \tag{9.34}$$

it can be rewritten as

$$\frac{d}{dt}\left[\mu\dot{u}\right] - \dot{\mu}\dot{u} + \Gamma(t)\mu\dot{u} + \mu\omega^2(t)u = 0\,. \tag{9.35}$$

We then choose $\dot{\mu} = \Gamma(t)\mu$, so that the integration factor is

$$\mu(t) = e^{\int^t \Gamma(\tau)d\tau}\,, \tag{9.36}$$

and the corresponding operator

$$\mathcal{L}_S \equiv \mu(t)\mathcal{L} = \mu(t)\frac{d^2}{dt^2} + \mu(t)\Gamma(t)\frac{d}{dt} + \mu(t)\omega^2(t) \tag{9.37}$$

is self-adjoint. The operator $\mathcal{L}_S = \mu(t)\mathcal{L}$, corresponding to differential equation (9.34), leads to the same equation of motion as $\mathcal{L}$, except that it is now self-adjoint.

Definition 9.2. The action corresponding to the self-adjoint linear operator $\mathcal{L}_S$ is

$$S = \frac{1}{2}\langle u|\mathcal{L}_S u\rangle = \frac{1}{2}\int_b^a u\left[\frac{d}{dt}(\mu\dot{u}) + \mu\omega^2(t)\right] u\, dt\,, \tag{9.38}$$

where the coefficient $\frac{1}{2}$ comes from the symmetry $\langle u|\mathcal{L}_S u\rangle = \langle \mathcal{L}_S u|u\rangle$.

From this definition,

$$S = \frac{1}{2}\langle \mathcal{L}_S u|u\rangle = \frac{1}{2}\int_b^a \left[-\mu\dot{u}^2 + \mu\omega^2(t)u^2\right] dt, \tag{9.39}$$

and by substitution of the integration factor

$$\begin{aligned} S &= \frac{1}{2}\langle u(t)|\mathcal{L}_S|u(t)\rangle \equiv \frac{1}{2}\langle u(t)|\mathcal{L}_S u(t)\rangle = \frac{1}{2}\int_b^a u(t)\mathcal{L}_S u(t)dt \\ &= \frac{1}{2}\int_b^a u\left[\frac{d}{dt}\left[e^{\int^t \Gamma(\tau)d\tau}\dot{u}\right] + e^{\int^t \Gamma(\tau)d\tau}\omega^2(t)u\right] dt \end{aligned} \tag{9.40}$$

we find

$$\begin{aligned} \frac{1}{2}\langle \mathcal{L}_S u(t)|u(t)\rangle &= \frac{1}{2}\int_b^a \mathcal{L}u(t)u(t)dt \\ &= \frac{1}{2}u\mu\dot{u}\Big|_{t=a}^{b} + \frac{1}{2}\int_b^a \left[-e^{\int^t \Gamma(\tau)}\dot{u}^2 + e^{\int^t \Gamma(\tau)d\tau}\omega^2(t)u^2\right] dt\,. \end{aligned}$$

From this action for the self-adjoint system we obtain the following Lagrangian function:

$$L = \frac{1}{2}e^{\int^t \Gamma(\tau)d\tau}\dot{u}^2 - \frac{1}{2}\omega^2(t)e^{\int^t \Gamma(\tau)d\tau}u^2\,, \tag{9.41}$$

the corresponding canonical momentum

$$p_u = \frac{\partial L}{\partial \dot{u}} = e^{\int^t \Gamma(\tau)d\tau}\dot{u} \tag{9.42}$$

and the Hamiltonian function

$$H(u, p_u) = \frac{1}{2}e^{-\int^t \Gamma(\tau)d\tau}{p_u}^2 + \frac{1}{2}e^{\int^t \Gamma(\tau)d\tau}\omega^2 u^2\,. \tag{9.43}$$

9.1.3 Particular cases of the self-adjoint equation

Case I: The damping coefficient Γ is a constant and the frequency $\omega(t)$ is the generic function of time. The integration factor then becomes explicitly $e^{\Gamma t}$. The Lagrangian function

$$L = \frac{1}{2}e^{\Gamma t}\dot{u}^2 - \frac{1}{2}\omega^2(t)e^{\Gamma t}u^2 \tag{9.44}$$

determines the equation of motion

$$\ddot{u} + \Gamma\dot{u} + \omega^2(t)u = 0 \tag{9.45}$$

and the generalized momentum

$$p_u = \frac{\partial L}{\partial \dot{u}} = e^{\Gamma t}\dot{u}\,. \tag{9.46}$$

Using the Legendre transformation, we get

$$H = \frac{1}{2}e^{\Gamma t}\dot{u}^2 + \frac{1}{2}\omega^2(t)u^2\,, \tag{9.47}$$

and after substituting for velocities $\dot{u}^2 = {p_u}^2 e^{-2\Gamma t}$, we find the Hamiltonian function

$$H(u, p_u) = e^{-\Gamma t}\frac{1}{2}{p_u}^2 + \frac{1}{2}e^{\Gamma t}\omega^2(t)u^2\,. \tag{9.48}$$

Case II: The frequency ω^2 is a constant, and damping is the generic function of time $\Gamma(t)$. The corresponding Lagrangian function

$$L = \frac{1}{2}e^{\int^t \Gamma(\tau)d\tau}\dot{u}^2 - \frac{1}{2}\omega^2 e^{\int^t \Gamma(\tau)d\tau}u^2\,, \tag{9.49}$$

leads to the following equation of motion:

$$\ddot{u} + \Gamma(t)\dot{u} + \omega^2 u = 0\,.$$

The corresponding generalized momentum

$$p_u = \frac{\partial L}{\partial \dot{u}} = e^{\int^t \Gamma(\tau)d\tau}\dot{u} \tag{9.50}$$

gives the Hamiltonian function

$$H(u, p_u) = e^{-\int^t \Gamma(\tau)d\tau}\frac{1}{2}{p_u}^2 + \frac{1}{2}e^{\int^t \Gamma(\tau)d\tau}\omega^2 u^2\,. \tag{9.51}$$

Case III: The damping coefficient Γ and the frequency ω^2 are constants. Then, the Lagrangian function

$$L = \frac{1}{2}e^{\Gamma t}\dot{u}^2 - \frac{1}{2}\omega^2 e^{\Gamma t}u^2 \tag{9.52}$$

gives the equation of motion

$$\ddot{u} + \Gamma \dot{u} + \omega^2 u = 0 \tag{9.53}$$

and the generalized momentum

$$p_u = \frac{\partial L}{\partial \dot{u}} = e^{\Gamma t} \dot{u} \,. \tag{9.54}$$

Using the Legendre transformation, we obtain the Hamiltonian function for the Caldirola–Kanai model,

$$H(u, p_u) = e^{-\Gamma t} \frac{1}{2} {p_u}^2 + \frac{1}{2} e^{\Gamma t} \omega^2 u^2 \,. \tag{9.55}$$

9.2 Oscillator equation with three regular singular points

As we have seen, every second-order homogeneous differential equation can be represented in oscillator form with damping and frequency depending on time functions. The singular points of these functions allows one to classify the equations and develop a series solution. In this subsection we consider the special case of second-order equations with three singular points. If regular singular points are fixed as 0, 1, and ∞, the solution of this equation is known as the Gauss hypergeometric function. When two singular points merge together, the equation take on the confluent hypergeometric form and includes, in particular cases, many special functions, such as the Bessel functions, the Hermite polynomials, the Laguerre polynomials, etc.

With this in mind, here we describe particular forms of variable damping and frequencies with fixed singularities and their corresponding solutions in terms of the special functions. In every case we provide the Lagrangian formulation for the nonself-adjoint form in terms of the doubled oscillator representation, as well as in the self-adjoint representation. Then, using these Lagrangians, we construct corresponding Hamiltonian descriptions. The canonical homogeneous second-order differential equation in self-adjoint form is

$$\frac{d}{dx}\left[p(x)\frac{du}{dx}\right] + [w(x)\lambda - q(x)]\, u(x) = 0 \,, \tag{9.56}$$

where we assume that $x \equiv t$. Then, equation (9.56) becomes

$$p(t)\frac{d^2 u(t)}{dt^2} + \dot{p}(t)\frac{du}{dt} + w(t)\lambda u(t) - q(t)u(t) = 0 \,, \tag{9.57}$$

where $p(t)$, $q(t)$ are the coefficient functions, λ is the eigenvalue and $w(t)$ is the weight function. The specific forms of these functions and eigenvalues, corresponding to different special functions, are given in Table 9.1.

Tab. 9.1: The coefficient functions and eigenvalues for the self-adjoint form of the special functions.

Equation	$p(t)$	$q(t)$	λ	$w(t)$
Bessel	t	n^2/t	1	t
Legendre	$1-t^2$	0	l(l+1)	1
Shifted Legendre	$t(1-t)$	0	l(l+1)	1
Associated Legendre	$1-t^2$	$m^2/(1-t^2)$	l(l+1)	1
Hermite	e^{-t^2}	0	2α	e^{-t^2}
Ultraspherical	$(1-t^2)^{\alpha+\frac{1}{2}}$	0	$n(n+2\alpha)$	$(1-t^2)^{\alpha-1/2}$
Laguerre	te^{-t}	0	α	e^{-t}
Associated Laguerre	$t^{k+1}e^{-t}$	0	$\alpha-k$	$t^k e^{-k}$
Chebyshev I	$(1-t^2)^{1/2}$	0	n^2	$(1-t^2)^{-1/2}$
Chebyshev II	$(1-t^2)^{3/2}$	0	$n(n+2)$	$(1-t^2)^{1/2}$
Shifted Chebyshev I	$[t(1-t)]^{1/2}$	0	n^2	$[t(1-t)]^{-1/2}$
Simple harmonic oscillator	1	0	n^2	1

We can represent this equation in oscillator form by dividing on $p(t)$,

$$\frac{d^2u}{dt^2} + \frac{\dot{p}(t)}{p(t)}\frac{du}{dt} + \left(\frac{q(t)}{p(t)} - \frac{w(t)}{p(t)}\lambda\right)u = 0, \tag{9.58}$$

and defining the corresponding damping and frequency variable parameters as

$$\Gamma(t) \equiv \frac{\dot{p(t)}}{p(t)}, \quad \omega^2(t) \equiv \left(\frac{q(t)}{p(t)} - \frac{w(t)}{p(t)}\lambda\right). \tag{9.59}$$

In addition to the damping and the frequency terms, we can obtain the integration factor as follows:

$$\mu(t) = e^{\int^t \Gamma(\tau)d\tau} = e^{\int^t \frac{\dot{p}}{p}d\tau} = p(t)\,. \tag{9.60}$$

Then, equation (9.57) is reduced to the following parametric oscillator form:

$$\ddot{u} + \Gamma(t)\dot{u} + \omega^2(t)u = 0\,. \tag{9.61}$$

In Table 9.2 we reproduce the damping $\Gamma(t)$ and frequency $\omega^2(t)$ functions related to several special functions. The second-order differential operator, corresponding to equation (9.61), is given by

$$\mathcal{L} = \frac{d^2}{dt^2} + \Gamma(t)\frac{d}{dt} + \omega^2(t)\,, \tag{9.62}$$

which is not a self-adjoint operator in general. However, by using the integration factor $\mu(t)$, as in (9.60), we can construct the self-adjoint operator. The orthogonality intervals $[a, b]$ for the special functions are given in Table 9.3.

After introducing the new time variable τ in equation (9.56) as

$$p(t)\frac{d}{dt} \equiv \frac{d}{d\tau} \quad \rightarrow \quad \frac{dt}{p(t)} = d\tau$$

Tab. 9.2: The damping and frequency functions corresponding to the special functions.

Equation	Damping $\Gamma(t)$	Frequency $\omega^2(t)$
Hypergeometric	$\frac{c-(a+b+1)t}{t(1-t)}$	$-\frac{ab}{t(1-t)}$
Confluent hypergeometric	$\frac{c-t}{t}$	$-\frac{a}{t}$
Bessel	$\frac{1}{t}$	$1-\frac{n^2}{t^2}$
Legendre	$-\frac{2t}{1-t^2}$	$\frac{l(l+1)}{1-t^2}$
Shifted Legendre	$\frac{1-2t}{t(1-t)}$	$\frac{l(l+1)}{t(1-t)}$
Associated Legendre	$-\frac{2t}{1-t^2}$	$\frac{1}{1-t^2}\left(l(l+1)-\frac{m^2}{1-t^2}\right)$
Hermite	$-2t$	2α
Ultraspherical	$-(2\alpha+1)\frac{t}{1-t^2}$	$\frac{n(n+2\alpha)}{1-t^2}$
Laguerre	$\frac{1-t}{t}$	$\frac{\alpha}{t}$
Associated Laguerre	$\frac{k+1-t}{t}$	$\frac{n}{t}$
Chebyshev I	$-\frac{t}{1-t^2}$	$\frac{n^2}{1-t^2}$
Chebyshev II	$-\frac{3t}{1-t^2}$	$\frac{n(n+2)}{1-t^2}$
Shifted Chebyshev I	$\frac{1}{2}\frac{1-2t}{t(1-t)}$	$\frac{n^2}{t(1-t)}$

Tab. 9.3: The orthogonality intervals $[a, b]$ for the special functions.

Equation	a	b
Hypergeometric	0	1
Confluent hypergeometric	0	∞
Bessel	0	∞
Legendre	-1	1
Shifted Legendre	0	1
Associated Legendre	-1	1
Hermite	$-\infty$	∞
Ultraspherical	-1	1
Laguerre	0	∞
Associated Laguerre	0	∞
Chebyshev I	-1	1
Chebyshev II	-1	1
Shifted Chebyshev I	0	1

and

$$\tau = \tau(t) = \int^{t} \frac{d\xi}{p(\xi)} ,$$

we obtain the harmonic oscillator equation

$$\ddot{u} + \Omega^2(\tau)u = 0 , \tag{9.63}$$

where the time dependent frequency is defined by formula

$$\Omega^2(t(\tau)) \equiv p(t(\tau))[q(t(\tau)) - \lambda w(t(\tau))] .$$

The new time variable τ can be written in terms of dissipation coefficient function $\Gamma(t)$ as

$$\tau = \int^{t} e^{\int^{\xi} \Gamma(\eta) d\eta} d\xi .$$

This implies that by choosing the proper time variable τ, damped oscillator equation (9.61) can be transformed into the purely time dependent frequency oscillator form (9.63).

Here we like to stress that in the general oscillatory representation of the Sturm–Liouville problem, the effective frequencies $\omega^2(t)$ and $\Omega^2(\tau)$ could change the sign at some values of the time. This means that in some time intervals our parametric harmonic oscillator may appear in a hyperbolic form and hence become unstable. It is an interesting problem to analyze cosmological models in such a type of oscillatory representation. In such models we can observe epochs with different behaviors: for the positive effective frequency, the oscillating character; and for the negative effective frequency, the hyperbolic dissipative (antidissipative) character. According to this we can describe the dynamics of a universe that includes both possibilities during its evolution.

Below we give the Lagrangian and the Hamiltonian description for the damped parametric oscillator representation corresponding to the main special functions of mathematical physics. In the nonself-adjoint case we get the doubled oscillator representation, while for the self-adjoint case no extension of the phase space is required. In the next chapter we quantize the damped harmonic oscillator with general time dependent effective frequency ω^2, when it is positive. It is characterized by a quasidiscrete energy spectrum with the wave function written in terms of Hermite polynomials with a time dependent argument. The quantization of a hyperbolic unstable oscillator, when effective frequency ω^2 is negative and no bound states or quasidiscrete states exist, is still an unsolved problem.

9.2.1 Hypergeometric oscillator

The hypergeometric differential equation

$$t(1-t)\,\ddot{u} + (c - (a+b+1)\,t)\,\dot{u} - abu = 0 \tag{9.64}$$

was introduced as a canonical form of a linear second-order differential equation with regular singularities at $t = 0, 1$ and ∞. The solution of this equation in the form of hypergeometric series converges for $|t| < 1$ and $t = 1$, for $c > a + b$, and $t = -1$, for $c > a + b - 1$.

After dividing equation (9.64) by $t(1-t)$, the parametric oscillator form is obtained:

$$\ddot{u} + \frac{(c - (a+b+1)\,t)}{t\,(1-t)}\dot{u} - \frac{ab}{t\,(1-t)}u = 0\,, \tag{9.65}$$

where the damping and the frequency functions are given by

$$\Gamma(t) \equiv \frac{[c - (a+b+1)\,t]}{t\,(1-t)}\,, \quad \omega^2(t) \equiv -\frac{ab}{t\,(1-t)}\,. \tag{9.66}$$

The oscillator representation of the hypergeometric differential equation then becomes

$$\ddot{u} + \Gamma(t)\dot{u} + \omega^2(t)u = 0\,. \tag{9.67}$$

Here the differential operator is not self-adjoint. For this nonself-adjoint operator, the two equations (9.17) and (9.18) give us the double oscillator representation in the form

$$\ddot{u} + \Gamma\dot{u} + \omega^2 u = 0\,,$$
$$\ddot{v} - \Gamma(t)\dot{v} + \left(\omega^2(t) - \dot{\Gamma}\right)v = 0\,.$$

By substituting the damping $\Gamma(t)$ and the frequency $\omega^2(t)$ functions, we get the following double oscillator representation of the hypergeometric equation:

$$\ddot{u} + \frac{[c - (a+b+1)\,t]}{t\,(1-t)}\dot{u} - \frac{ab}{t\,(1-t)}u = 0\,, \tag{9.68}$$

$$\ddot{v} + \frac{(a+b+1)\,t - c}{t\,(1-t)}\dot{v} + \left(\frac{c - 2ct + t(-ab + (1+a)(1+b)t)}{t^2(t-1)^2}\right)v = 0\,. \tag{9.69}$$

The solution of oscillator (9.68) in the series form

$$u(t) = {}_2F_1(a,b,c;t) = 1 + \frac{ab}{c}\frac{t}{1!} + \frac{a(a+1)b(b+1)}{c(c+1)}\frac{t^2}{2!} + \ldots = \sum_{n=0}^{\infty}\frac{(a)_n(b)_n}{(c)_n}\frac{t^n}{n!}$$

represents the Gauss hypergeometric function. The Pochhammer symbol is definite here as $(a)_0 = 1$, and

$$(a)_n = a(a+1)(a+2)\ldots(a+n-1) = \frac{(a+n-1)!}{(a-1)!}\,.$$

The Lagrangian function corresponding to this doubled system of equations is the following:

$$L = \dot{u}\dot{v} - \frac{[c - (a+b+1)\,t]}{2t\,(1-t)}(v\dot{u} - \dot{v}u)$$
$$- \left(\frac{c - 2(ab+c)t + (1+a+b+2ab)t^2}{2t^2(t-1)^2}\right)uv\,. \tag{9.70}$$

From this Lagrangian we find the momenta of the doubled system as

$$p_u = \frac{\partial L}{\partial \dot{u}} = \dot{v} - \frac{1}{2}\frac{[c-(a+b+1)\,t]}{t\,(1-t)}v\,, \qquad p_v = \frac{\partial L}{\partial \dot{v}} = \dot{u} + \frac{1}{2}\frac{[c-(a+b+1)\,t]}{t\,(1-t)}u\,.$$

Finally, by the Legendre transformation we obtain the Hamiltonian function for the nonself-adjoint case of the damped hypergeometric oscillator

$$\begin{aligned} H(u,v,p_u,p_v) &= p_u p_v + \frac{1}{2}\frac{[c-(a+b+1)\,t]}{t\,(1-t)}(vp_v - up_u) + \\ &+ \left(\frac{-c^2 + t(-4ab + t - (a-b)^2 t) + 2c(1+(-1+a+b)t)}{4t^2(t-1)^2}\right)uv\,. \end{aligned} \tag{9.71}$$

It is also possible to obtain the self-adjoint form of the damped hypergeometric oscillator equation (9.67) by multiplying it with an integration factor. The integration factor for equation (9.67), to transform it into the self-adjoint form, is

$$\mu = e^{\int^t \frac{c-(a+b+1)\tau}{\tau(1-\tau)}d\tau} = t^c(1-t)^{1+a+b-c}\,. \tag{9.72}$$

The Lagrangian function in the self-adjoint case takes the form

$$L = \frac{t^c(1-t)^{1+a+b-c}}{2}\left(\dot{u}^2 + \frac{ab}{t(1-t)}u^2\right) \tag{9.73}$$

and determines the momentum

$$p_u = t^c(1-t)^{1+a+b-c}\dot{u}\,. \tag{9.74}$$

Then, for the self-adjoint case we obtain the Hamiltonian of the hypergeometric damped oscillator as

$$H(u,p_u) = \frac{(1-t)^{1+a+b}t^c}{2(1-t)^c}\left(\frac{(1-t)^{2c}{p_u}^2}{t^{2c}(1-t)^{2(1+a+b)}} - \frac{ab}{t(1-t)}u^2\right). \tag{9.75}$$

9.2.2 Confluent hypergeometric oscillator

The confluent hypergeometric equation

$$t\ddot{u} + (c-t)\,\dot{u} - au = 0 \tag{9.76}$$

may be obtained from the hypergeometric equation by merging two of its singularities. The equation has a regular singularity at $t = 0$ and an irregular one at $t = \infty$. After dividing equation (9.76) by t,

$$\ddot{u} + \frac{c-t}{t}\dot{u} - \frac{a}{t}u = 0\,, \tag{9.77}$$

we obtain the confluent hypergeometric damped harmonic oscillator equation

$$\ddot{u} + \Gamma(t)\dot{u} + \omega^2(t)u = 0\,,$$

with time dependent damping and frequency of the form

$$\Gamma(t) \equiv \frac{c-t}{t}\,, \quad \omega^2(t) \equiv -\frac{a}{t}\,.$$

The equation is not self-adjoint. This is why we form the doubled oscillator representation for this equation as

$$\ddot{u} + \frac{c-t}{t}\dot{u} - \frac{a}{t}u = 0\,, \tag{9.78}$$

$$\ddot{v} - \frac{c-t}{t}\dot{v} + \left(\frac{c}{t^2} - \frac{a}{t}\right)v = 0\,. \tag{9.79}$$

The following Lagrangian corresponds to this system of double oscillators:

$$L = \dot{u}\dot{v} - \frac{1}{2}\frac{(c-t)}{t}(v\dot{u} - \dot{v}u) - \left(\frac{1}{2}\frac{c}{t^2} - \frac{a}{t}\right)uv \tag{9.80}$$

and determines the momenta of the confluent hypergeometric doublet system as

$$p_u = \frac{\partial L}{\partial \dot{u}} = \dot{v} - \frac{1}{2}\frac{(c-t)}{t}v\,, \quad p_v = \frac{\partial L}{\partial \dot{v}} = \dot{u} + \frac{1}{2}\frac{(c-t)}{t}u\,. \tag{9.81}$$

Finally, we can write the Hamiltonian function as

$$H(u, v, p_u, p_v) = p_u p_v + \frac{(c-t)}{2t}(vp_v - up_u) + \left(\frac{c}{2t^2}\left(1 - \frac{c}{2}\right) + \frac{1}{t}\left(\frac{c}{2} - a\right) - \frac{1}{4}\right)uv\,.$$

We can also construct a self-adjoint differential operator for the damped confluent hypergeometric oscillator. The integration factor we obtain in the form

$$\mu(t) = \frac{t^c}{e^t}\,, \tag{9.82}$$

and the Lagrangian function for this self-adjoint case is

$$L = \frac{t^c}{2e^t}\left(\dot{u}^2 + \frac{a}{t}u^2\right)\,. \tag{9.83}$$

After deriving the momentum,

$$p_u = \frac{t^c}{e^t}\dot{u}\,, \tag{9.84}$$

we find the Hamiltonian for the self-adjoint damped confluent hypergeometric oscillator as

$$H(u, p_u) = \frac{t^c}{2e^t}\left(\frac{e^{2t}}{t^{2c}}{p_u}^2 - \frac{a}{t}u^2\right)\,. \tag{9.85}$$

9.2.3 Bessel oscillator

The Bessel differential equation is

$$t^2\ddot{u} + t\dot{u} + \left(t^2 - n^2\right)u = 0\,. \tag{9.86}$$

After dividing this equation by t, we obtain the Bessel damped oscillator

$$\ddot{u} + \frac{1}{t}\dot{u} + \left(1 - \frac{n^2}{t^2}\right)u = 0\,, \tag{9.87}$$

where the damping and the frequency are

$$\Gamma \equiv \frac{1}{t}, \qquad \omega^2 \equiv 1 - \frac{n^2}{t^2}\,.$$

To construct the Lagrangian function for this nonself-adjoint equation, we extend the phase space to the double oscillator equations

$$\ddot{u} + \frac{1}{t}\dot{u} + \left(1 - \frac{n^2}{t^2}\right)u(t) = 0\,,$$

$$\ddot{v} - \frac{1}{t}\dot{v} + \frac{1}{t^2}\left(1 - n^2 + t^2\right)v = 0\,. \tag{9.88}$$

Then, the Lagrangian of this doubled system of the Bessel differential equations is

$$L = \dot{u}\dot{v} - \frac{1}{2t}\left(v\dot{u} - \dot{v}u\right) - \frac{1}{t^2}\left(\frac{1}{2} - n^2 + t^2\right)uv\,. \tag{9.89}$$

Defining the generalized momenta as

$$p_u = \frac{\partial L}{\partial \dot{u}} = \dot{v} - \frac{1}{2t}v\,, \quad p_v = \frac{\partial L}{\partial \dot{v}} = \dot{u} + \frac{1}{2t}u\,, \tag{9.90}$$

and by using the Legendre transformation, we find the Hamiltonian of the Bessel damped double oscillator system as

$$\begin{aligned} H(u, v, p_u, p_v) = p_u p_v &+ \frac{1}{2t}\left(vp_v - up_u\right) \\ &+ \frac{1}{t^2}\left(\frac{1}{4} - n^2 + t^2\right)uv\,. \end{aligned} \tag{9.91}$$

It is possible to get the self-adjoint form of equation (9.87) by multiplying it with an integration factor. The integration factor of this equation is

$$\mu(t) = t\,, \tag{9.92}$$

and the corresponding Lagrangian of the Bessel damped oscillator for the self-adjoint case is defined as

$$L = \frac{t}{2}\left(\dot{u}^2 - \left(1 - \frac{n^2}{t}\right)u^2\right)\,. \tag{9.93}$$

For the momenta we find

$$p_u = t\dot{u}\,, \tag{9.94}$$

and the Hamiltonian for the self-adjoint Bessel damped oscillator is

$$H(u, p_u) = \frac{t}{2}\left(\frac{p_u^2}{t^2} + \left(1 - \frac{n^2}{t}\right)u^2\right)\,. \tag{9.95}$$

9.2.4 Legendre oscillator

From the Legendre differential equation

$$(1 - t^2)\ddot{u} - 2t\dot{u} + l(l+1)u = 0\,,$$

we obtain the time dependent Legendre-type oscillator equation

$$\ddot{u} - \frac{2t}{1-t^2}\dot{u} + \frac{l(l+1)}{1-t^2}u = 0\,, \tag{9.96}$$

where the damping and the frequency functions take the form

$$\Gamma(t) \equiv -\frac{2t}{1-t^2}\,, \qquad \omega^2(t) \equiv \frac{l(l+1)}{1-t^2}\,.$$

As we can easily see, this equation has the nonself-adjoint form. Therefore, we have the doubled Legendre damped oscillator as the following system:

$$\ddot{u} - \frac{2t}{1-t^2}\dot{u} + \frac{l(l+1)}{1-t^2}u = 0\,, \tag{9.97}$$

$$\ddot{v} + \frac{2t}{1-t^2}\dot{v} - \frac{1}{1-t^2}\left(2 + l(l+1) + \frac{4t^2}{(1-t^2)}\right)v = 0\,. \tag{9.98}$$

The solutions of (9.97) are given by the Legendre polynomials

$$u(t) = P_l(t) = \sum_{k=0}^{[l/2]} (-1)^k \frac{(2l-2k)!}{2^l k!(l-k)!(l-2k)!} t^{l-2k}\,.$$

The Lagrangian of this doubled system is

$$L = \dot{u}\dot{v} + \frac{t}{1-t^2}(v\dot{u} - \dot{v}u) - \frac{1}{1-t^2}\left(1 + l(l+1) + \frac{2t^2}{(1-t^2)}\right)uv\,. \tag{9.99}$$

Using this Lagrangian, the momenta of the system are obtained as

$$p_u = \dot{v} + \frac{t}{1-t^2}v\,, \qquad p_v = \dot{u} - \frac{t}{1-t^2}u\,. \tag{9.100}$$

Then, the Hamiltonian of the doubled Legendre damped oscillator system is found as

$$H(u, v, p_u, p_v) = p_u p_v - \frac{t}{1-t^2}(v p_v - u p_u) + \frac{1}{1-t^2}\left(1 + l(l+1) + \frac{t^2}{1-t^2}\right)uv\,.$$

To get the self-adjoint form of the Legendre oscillator (9.96), we find the integration factor as

$$\mu(t) = t^2 - 1\,. \tag{9.101}$$

The Lagrangian function for the self-adjoint case is

$$L = \frac{t^2-1}{2}\left(\dot{u}^2 - \frac{l(l+1)}{1-t^2}u^2\right)\,, \tag{9.102}$$

and the corresponding momentum defined as

$$p_u = \left(t^2 - 1\right)\dot{u}\,, \tag{9.103}$$

which finally gives us the Hamiltonian of the Legendre damped oscillator,

$$H(u, p_u) = \frac{t^2 - 1}{2}\left(\frac{{p_u}^2}{(t^2 - 1)^2} + \frac{l(l+1)}{1 - t^2}u^2\right). \tag{9.104}$$

9.2.5 Shifted Legendre oscillator

The shifted Legendre differential equation is

$$t(1 - t)\ddot{u} + (1 - 2t)\dot{u} + l(l+1)u = 0. \tag{9.105}$$

Dividing it by $t(1 - t)$ we obtain the damped oscillator

$$\ddot{u} + \frac{1 - 2t}{t(1 - t)}\dot{u} + \frac{l(l+1)}{t(1 - t)}u = 0\,, \tag{9.106}$$

where the damping term is $\Gamma(t) \equiv \frac{1-2t}{t(1-t)}$, while the frequency is $\omega^2(t) \equiv \frac{l(l+1)}{t(1-t)}$. Due to the nonself-adjoint form of the equation, we can construct the doubled oscillator representation of the shifted Legendre equation as follows:

$$\ddot{u} + \frac{1 - 2t}{t(1 - t)}\dot{u} + \frac{l(l+1)}{t(1 - t)}u = 0\,, \tag{9.107}$$

$$\ddot{v} - \frac{1 - 2t}{t(1 - t)}\dot{v} + \left(\frac{1}{(t-1)^2} + \frac{1}{t^2} + \frac{l(l+1)+1}{t(t-1)}\right)v = 0\,. \tag{9.108}$$

The Lagrangian function corresponding to this system is

$$L = \dot{u}\dot{v} - \frac{1 - 2t}{2t(1 - t)}(v\dot{u} - \dot{v}u) - \left[\frac{1 + l(l+1)}{t(1 - t)}\frac{1}{2}\left(\frac{1}{(t-1)^2} + \frac{1}{t^2}\right)\right]uv\,, \tag{9.109}$$

the momenta are

$$p_u = \dot{v} - \frac{1 - 2t}{2t(1 - t)}v\,, \quad p_v = \dot{u} + \frac{1 - 2t}{2t(1 - t)}u\,, \tag{9.110}$$

and the Hamiltonian function of the shifted Legendre doubled oscillator is

$$\begin{aligned} H(u, v, p_u, p_v) &= p_u p_v + \frac{1}{2}\frac{1 - 2t}{t(1 - t)}(vp_v - up_u) \\ &+ \frac{1}{t(1 - t)}\left(l(l+1) + 1 + \frac{1 + 2t(t-2)}{2t(t-1)}\right)uv\,. \end{aligned} \tag{9.111}$$

In addition to this doubled oscillator representation, we can obtain the integration factor of equation (9.106) as

$$\mu = t\,(1 - t) \tag{9.112}$$

and construct the self-adjoint form of the shifted Legendre oscillator equation.

The Lagrangian function for this self-adjoint case is given by

$$L = \frac{t\,(1-t)}{2}\left(\dot{u}^2 - \frac{l(l+1)}{t(1-t)}u^2\right). \tag{9.113}$$

The corresponding momentum is

$$p_u = t\,(1-t)\,\dot{u}\,. \tag{9.114}$$

Finally, the Hamiltonian function for the self-adjoint shifted Legendre damped oscillator we obtain as

$$H(u, p_u) = \frac{t\,(1-t)}{2}\left(\frac{p_u{}^2}{t^2(t-1)^2} + \frac{l(l+1)}{t(1-t)}u^2\right). \tag{9.115}$$

9.2.6 Associated Legendre oscillator

The associated Legendre differential equation with time as an independent variable is the following:

$$(1-t^2)\ddot{u} - 2t\dot{u} + \left[l(l+1) - \frac{m^2}{1-t^2}\right]u = 0\,. \tag{9.116}$$

Dividing it by $1 - t^2$ we obtain the associated Legendre oscillator

$$\ddot{u} - \frac{2t}{1-t^2}\dot{u} + \frac{1}{1-t^2}\left[l(l+1) - \frac{m^2}{1-t^2}\right]u = 0, \tag{9.117}$$

in which the damping and the frequency terms are given as $\Gamma(t) \equiv -\frac{2t}{1-t^2}$ and $\omega^2(t) \equiv \frac{1}{1-t^2}\left(l(l+1) - \frac{m^2}{1-t^2}\right)$. Note that the associated Legendre equation has nonzero solutions, which are nonsingular on $t \in [-1, 1]$ as long as l and m are integers, satisfying $0 \le m \le l$, or with trivially equivalent negative values. In addition to this condition, if m is even, then the solution is given by a polynomial. When m is zero and l is an integer, the solutions satisfying these conditions are identical to the Legendre polynomials. The doubled oscillator representation of the associated Legendre equation is

$$\begin{aligned} \ddot{u} - \frac{2t}{1-t^2}\dot{u} + \frac{1}{1-t^2}\left[l(l+1) - \frac{m^2}{1-t^2}\right]u = 0\,, \\ \ddot{v} + \frac{2t}{1-t^2}\dot{v} + \left(-\frac{1}{1-t^2}\left[\frac{m^2}{1-t^2} + l(l+1)\right] + \frac{2\,(1+t^2)}{(1-t^2)^2}\right)v = 0\,. \end{aligned} \tag{9.118}$$

We find the corresponding Lagrangian function

$$L = \dot{u}\dot{v} + \frac{1}{2}\frac{2t}{1-t^2}\,(v\dot{u} - \dot{v}u) - \frac{1}{1-t^2}\left(l(l+1) + 1 + \frac{2t^2 - m^2}{1-t^2}\right)uv\,, \tag{9.119}$$

generalized momenta

$$p_u = \dot{v} + \frac{t}{1-t^2} v , \quad p_v = \dot{u} - \frac{t}{1-t^2} u , \tag{9.120}$$

and the Hamiltonian

$$H(u, v, p_u, p_v) = p_u p_v - \frac{t}{1-t^2}(v p_v - u p_u) + \frac{1}{1-t^2}\left(l(l+1) + 1 + \frac{t^2 - m^2}{1-t^2}\right) uv .$$

To get the self-adjoint form of equation (10.22) we find the integration factor as

$$\mu(t) = t^2 - 1 . \tag{9.121}$$

The Lagrangian equation of the damped harmonic oscillator of the associated Legendre function for the self-adjoint case is

$$L = \frac{t^2 - 1}{2}\left[\dot{u}^2 - \frac{1}{1-t^2}\left(l(l+1) - \frac{m^2}{1-t^2}\right) u^2\right] . \tag{9.122}$$

Then, by the generalized momentum

$$p_u = \left(t^2 - 1\right) \dot{u} , \tag{9.123}$$

we find the Hamiltonian function as

$$H(u, p_u) = -\frac{1}{2}\left[\frac{{p_u}^2}{(t^2-1)} + \left(l(l+1) - \frac{m^2}{1-t^2}\right) u^2\right] . \tag{9.124}$$

9.2.7 Hermite oscillator

The Hermite differential equation and corresponding Hermite oscillator is defined as

$$\ddot{u} - 2t\dot{u} + 2\alpha u = 0 . \tag{9.125}$$

In contrast to previous examples, this oscillator has the time dependent damping

$$\Gamma(t) \equiv -2t ,$$

while its frequency is a constant $\omega^2 \equiv 2\alpha$. Since the problem is not self-adjoint, we have the doubled oscillator representation for this model as a system

$$\begin{aligned} &\ddot{u} - 2t\dot{u} + 2\alpha u = 0 , \\ &\ddot{v} + 2t\dot{v} + 2(\alpha + 1) v = 0 . \end{aligned} \tag{9.126}$$

This double oscillator is determined by the following Lagrangian:

$$L = \dot{u}\dot{v} + t(v\dot{u} - \dot{v}u) - (2\alpha + 1) uv . \tag{9.127}$$

The generalized momenta of this doublet Hermite oscillatory system are

$$p_u = \dot{v} + tv\,, \quad p_v = \dot{u} - tu\,, \tag{9.128}$$

and the Hamiltonian function we obtain in the following form:

$$H(u, v, p_u, p_v) = p_u p_v - t\,(vp_v - up_u) + \left(2\alpha + 1 - t^2\right) uv\,. \tag{9.129}$$

By using the integration factor, we can get the self-adjoint form of the Hermite damped oscillator. As easy way to find the integration factor of equation (9.125) is

$$\mu(t) = e^{-\int^t 2\tau d\tau} = e^{-t^2}\,. \tag{9.130}$$

Thus, we can write the Lagrangian for the self-adjoint form of the Hermite damped oscillator as

$$L = \frac{e^{-t^2}}{2}\left(\dot{u}^2 - \omega^2 u^2\right)\,. \tag{9.131}$$

This gives the corresponding momentum

$$p_u = e^{-t^2}\dot{u} \tag{9.132}$$

and the self-adjoint Hamiltonian

$$H(u, p_u) = \frac{e^{-t^2}}{2}\left(e^{2t^2}{p_u}^2 + \omega^2 u^2\right)\,. \tag{9.133}$$

9.2.8 Gegenbauer ultraspherical oscillator

The ultraspherical (Gegenbauer) polynomials satisfy the following second-order differential equation:

$$(1 - t^2)^{\alpha+\frac{1}{2}}\ddot{u} - (2\alpha + 1)\,t(1 - t^2)^{\alpha-\frac{1}{2}}\dot{u} + n(n + 2\alpha)(1 - t^2)^{\alpha-\frac{1}{2}}u = 0\,. \tag{9.134}$$

By dividing this on $(1 - t^2)^{\alpha+\frac{1}{2}}$ we obtain the Gegenbauer damped oscillator

$$\ddot{u} - (2\alpha + 1)\,\frac{t}{1 - t^2}\dot{u} + \frac{n(n + 2\alpha)}{1 - t^2}u = 0\,, \tag{9.135}$$

with time dependent damping

$$\Gamma \equiv -(2\alpha + 1)\,\frac{t}{1 - t^2}$$

and the time dependent frequency

$$\omega^2 \equiv \frac{n(n + 2\alpha)}{1 - t^2}\,.$$

The double oscillator representation of this equation is given by the system

$$\ddot{u} - (2\alpha+1)\frac{t}{1-t^2}\dot{u} + \frac{n(n+2\alpha)}{1-t^2}u = 0\,, \tag{9.136}$$

$$\ddot{v} + (2\alpha+1)\frac{t}{1-t^2}\dot{v} + \frac{1}{1-t^2}\left(n(n+2\alpha) + (2\alpha+1)\frac{1+t^2}{1-t^2}\right)v = 0\,. \tag{9.137}$$

The Lagrangian for this doubled ultraspherical-type oscillator equation then becomes

$$L = \dot{u}\dot{v} + \left(\alpha+\frac{1}{2}\right)\frac{t}{1-t^2}(v\dot{u} - \dot{v}u) - \frac{1}{(1-t^2)}\left(n(n+2\alpha) + \left(\alpha+\frac{1}{2}\right)\frac{1+t^2}{1-t^2}\right)uv\,.$$

The corresponding momenta are

$$p_u = \dot{v} + \left(\alpha+\frac{1}{2}\right)\frac{t}{(1-t^2)}v\,, \quad p_v = \dot{u} - \left(\alpha+\frac{1}{2}\right)\frac{t}{(1-t^2)}u\,, \tag{9.138}$$

and the Hamiltonian function of the ultraspherical doublet system is obtained as

$$\begin{aligned} H(u,v,p_u,p_v) &= p_u p_v - \left(\alpha+\frac{1}{2}\right)\frac{t}{1-t^2}(vp_v - up_u) + \\ &+ \frac{1}{1-t^2}\left[n(n+2\alpha) + \frac{\left(\alpha+\frac{1}{2}\right)}{1-t^2}\left(1+t^2 - \left(\alpha+\frac{1}{2}\right)t\right)\right]uv\,. \end{aligned} \tag{9.139}$$

By using integration factor, we can get the self-adjoint form of the ultraspherical oscillator. The integration factor of equation (9.135) is

$$\mu = \left(t^2-1\right)^{\alpha+\frac{1}{2}} \tag{9.140}$$

and the Lagrangian function of the damped harmonic oscillator of ultraspherical (Gegenbauer) type appears as

$$L = \frac{\left(t^2-1\right)^{\alpha+\frac{1}{2}}}{2}\left(\dot{u}^2 - \frac{n(n+2\alpha)}{1-t^2}u^2\right)\,. \tag{9.141}$$

Then, by determining the corresponding momentum as

$$p_u = \left(t^2-1\right)^{\alpha+\frac{1}{2}}\dot{u}\,, \tag{9.142}$$

we obtain the Hamiltonian function for the self-adjoint case,

$$H(u,p_u) - \frac{\left(t^2-1\right)^{\alpha+\frac{1}{2}}}{2}\left(\frac{{p_u}^2}{\left(t^2-1\right)^{2\alpha+1}} + \frac{n(n+2\alpha)}{(1-t^2)}u^2\right)\,. \tag{9.143}$$

9.2.9 Laguerre oscillator

Dividing the Laguerre equation

$$t\ddot{u} + (1-t)\dot{u} + \alpha u = 0\,, \tag{9.144}$$

to t, gives the Laguerre oscillator

$$\ddot{u} + \frac{1-t}{t}\dot{u} + \frac{\alpha}{t}u = 0, \tag{9.145}$$

with the damping term and frequency given by functions

$$\Gamma(t) \equiv \frac{1-t}{t}, \qquad \omega^2(t) \equiv \frac{\alpha}{t}.$$

For this nonself-adjoint form we get the double oscillator representation

$$\ddot{u} + \frac{1-t}{t}\dot{u} + \frac{\alpha}{t}u = 0\,, \tag{9.146}$$

$$\ddot{v} - \frac{(1-t)}{t}\dot{v} + \frac{1}{t}\left(\alpha + \frac{1}{t}\right)v = 0\,. \tag{9.147}$$

The Lagrangian of the corresponding doublet representation of the time dependent Laguerre oscillator is

$$L = \dot{u}\dot{v} - \frac{(1-t)}{2t}(v\dot{u} - \dot{v}u) - \frac{1}{t}\left(\alpha + \frac{1}{2t}\right)uv\,. \tag{9.148}$$

Then, the momenta are

$$p_u = \dot{v} - \frac{(1-t)}{2t}v\,, \qquad p_v = \dot{u} + \frac{(1-t)}{2t}u\,, \tag{9.149}$$

and by using the Legendre transformation, the Hamiltonian of the Laguerre double oscillator is found as

$$H(u, v, p_u, p_v) = p_u p_v + \frac{(1-t)}{2t}(v p_v - u p_u) + \frac{1}{t}\left[\alpha + \frac{1}{2t}\left(1 - \frac{(1-t)^2}{2}\right)\right]uv\,.$$

In addition to the doubled oscillator representation, we can construct the self-adjoint form of equation (9.146) by using the integration factor. Here the integration factor is

$$\mu(t) = \frac{t}{e^t}\,. \tag{9.150}$$

The self-adjoint Lagrangian function of the damped Laguerre oscillator is given as

$$L = \frac{t}{2e^t}\left(\dot{u}^2 - \frac{\alpha}{t}u^2\right), \tag{9.151}$$

the corresponding generalized momentum is

$$p_u = \frac{t}{e^t}\dot{u}\,, \tag{9.152}$$

and the Hamiltonian we find as

$$H(u, p_u) = \frac{t}{2e^t}\left(\frac{e^{2t}}{t^2}{p_u}^2 + \frac{\alpha}{t}u^2u^2\right). \tag{9.153}$$

9.2.10 Associated Laguerre oscillator

The associated Laguerre equation

$$t\ddot{u} + (k+1-t)\dot{u} + (\alpha - k)u = 0 \tag{9.154}$$

can be rewritten as the time dependent damped oscillator – the associated Laguerre oscillator:

$$\ddot{u} + \frac{(k+1-t)}{t}\dot{u} + \frac{n}{t}u = 0, \tag{9.155}$$

where $n = \alpha - k$, and the damping parameter and the frequency are

$$\Gamma(t) \equiv \frac{(k+1-t)}{t}, \qquad \omega^2(t) \equiv \frac{n}{t}.$$

The doubled oscillator representation of the associated Laguerre differential equation is

$$\ddot{u} + \frac{(k+1-t)}{t}\dot{u} + \frac{n}{t}u = 0\,, \tag{9.156}$$

$$\ddot{v} - \frac{(k+1-t)}{t}\dot{v} + \frac{1}{t}\left(n+1+\frac{k}{t}\right)v = 0\,, \tag{9.157}$$

with corresponding Lagrangian function

$$L = \dot{u}\dot{v} - \frac{(k+1-t)}{2t}(v\dot{u} - \dot{v}u) - \frac{1}{t}\left(n+\frac{1}{2}+\frac{k}{t}\right)uv\,. \tag{9.158}$$

The generalized momenta and the Hamiltonian function in the dual formalism are correspondingly

$$p_u = \dot{v} - \frac{(k+1-t)}{2t}v\,, \qquad p_v = \dot{u} + \frac{(k+1-t)}{2t}u\,, \tag{9.159}$$

and

$$H = p_u p_v + \frac{(k+1-t)}{2t}(vp_v - up_u) + \frac{1}{t}\left[n+\frac{1}{2}+\frac{1}{2t}\left(k - \frac{(k+1-t)^2}{2}\right)\right]uv\,.$$

To transform the nonself-adjoint oscillator form into the self-adjoint one, we need an integration factor, which is

$$\mu = \frac{t^{k+1}}{e^t}\,. \tag{9.160}$$

The Lagrangian function for the self-adjoint case,

$$L = \frac{t^{k+1}}{2e^t}\left(\dot{u}^2 - \frac{n}{t}u^2\right), \tag{9.161}$$

gives the generalized momentum

$$p_u = \frac{t^{k+1}}{e^t}\dot{u} \tag{9.162}$$

and the Hamiltonian function for the associated Laguerre oscillator

$$H(u, p_u) = \frac{t^{k+1}}{2e^t}\left(\frac{e^{2t}}{t^{2(k+1)}}{p_u}^2 + \frac{n}{t}u^2\right). \tag{9.163}$$

9.2.11 Chebyshev I oscillator

The first kind Chebyshev differential equation is

$$(1-t^2)^{\frac{1}{2}}\ddot{u} - \frac{t}{\sqrt{1-t^2}}\dot{u} + (1-t^2)^{-\frac{1}{2}}n^2 u = 0. \tag{9.164}$$

After diving it by $(1-t^2)^{1/2}$, it is reduced to the Chebyshev I damped harmonic oscillator,

$$\ddot{u} - \frac{t}{1-t^2}\dot{u} + \frac{n^2}{1-t^2}u = 0, \tag{9.165}$$

where the damping and frequency terms depend on time as

$$\Gamma(t) \equiv -\frac{t}{1-t^2}, \qquad \omega^2(t) \equiv \frac{n^2}{1-t^2}.$$

Since in this form the damped oscillator equation is not self-adjoint, it is represented as the doubled oscillator system

$$\ddot{u} - \frac{t}{1-t^2}\dot{u} + \frac{n^2}{1-t^2}u = 0\,, \tag{9.166}$$

$$\ddot{v} + \frac{t}{1-t^2}\dot{v} + \frac{1}{1-t^2}\left(n^2 + \frac{1+t^2}{1-t^2}\right)v = 0\,. \tag{9.167}$$

The Lagrangian function for this, the first kind of Chebyshev double oscillator system, is

$$L = \dot{u}\dot{v} + \frac{t}{2\,(1-t^2)}\,(v\dot{u} - \dot{v}u) - \frac{1}{1-t^2}\left(n^2 + \frac{1+t^2}{2\,/1-t^2)}\right)uv\,. \tag{9.168}$$

The generalized momenta determined by this Lagrangian,

$$p_u = \dot{v} + \frac{t}{2\,(1-t^2)}v\,, \qquad p_v = \dot{u} - \frac{t}{2\,(1-t^2)}u\,, \tag{9.169}$$

give us the Hamiltonian of the Chebyshev I double oscillator as

$$\begin{aligned} H(u, v, p_u, p_v) &= p_u p_v - \frac{t}{2\,(1-t^2)}\,(vp_v - up_u) \\ &\quad + \frac{1}{1-t^2}\left[n^2 + \frac{1}{2\,/1-t^2)}\left(\frac{t^2}{2}+1\right)\right]uv\,. \end{aligned} \tag{9.170}$$

By using integration factor, we can get the self-adjoint form of the first kind of Chebyshev damped oscillator. The integration factor is

$$\mu = \sqrt{t^2-1} \tag{9.171}$$

and the Lagrangian function for the self-adjoint case is given as

$$L = \frac{\sqrt{t^2-1}}{2}\left(\dot{u}^2 - \frac{n^2}{1-t^2}u^2\right)\,. \tag{9.172}$$

By calculating the generalized momentum,

$$p_u = \sqrt{t^2 - 1}\dot{u}\,, \tag{9.173}$$

we find the Hamiltonian of the self-adjoint first kind of Chebyshev damped oscillator,

$$H(u, p_u) = \frac{\sqrt{t^2 - 1}}{2}\left(\frac{{p_u}^2}{t^2 - 1} + \frac{n^2}{1 - t^2}u^2\right). \tag{9.174}$$

9.2.12 Chebyshev II oscillator

The second kind of Chebyshev differential equation,

$$(1 - t^2)^{\frac{3}{2}}\ddot{u} - 3t(1 - t^2)^{\frac{1}{2}}\dot{u} + n(n + 2)(1 - t^2)^{\frac{1}{2}}u = 0\,, \tag{9.175}$$

divided by $(1 - t^2)^{\frac{3}{2}}$ is reduced to the damped Chebyshev II oscillator,

$$\ddot{u} - \frac{3t}{1 - t^2}\dot{u} + \frac{n(n + 2)}{1 - t^2}u = 0\,, \tag{9.176}$$

where time dependent damping and frequency terms are correspondingly

$$\Gamma(t) \equiv -\frac{3t}{1 - t^2}, \quad \omega^2(t) \equiv \frac{n(n + 2)}{1 - t^2}\,.$$

This form of the oscillator is not self-adjoint, so that we have it in the double oscillator representation as

$$\ddot{u} - \frac{3t}{1 - t^2}\dot{u} + \frac{n(n + 2)}{1 - t^2}u = 0\,, \tag{9.177}$$

$$\ddot{v} + \frac{3t}{1 - t^2}\dot{v} + \frac{1}{1 - t^2}\left(n(n + 2) + 3\frac{(1 + t^2)}{1 - t^2}\right)v = 0\,. \tag{9.178}$$

This double oscillator representation of the second kind of Chebyshev oscillator is determined by the following Lagrangian:

$$L = \dot{u}\dot{v} + \frac{3t}{2\,(1 - t^2)}\,(v\dot{u} - \dot{v}u) - \frac{1}{1 - t^2}\left(n(n + 2) + \frac{3}{2}\frac{(1 + t^2)}{1 - t^2}\right)uv\,. \tag{9.179}$$

Then, from the generalized momenta

$$p_u = \dot{v} + \frac{3l}{2\,(1 - t^2)}v\,, \quad p_v = \dot{u} - \frac{3t}{2\,(1 - t^2)}u\,, \tag{9.180}$$

we find the Hamiltonian function as

$$\begin{aligned} H(u, v, p_u, p_v) &= p_u p_v - \frac{3t}{2\,(1 - t^2)}\,(v p_v - u p_u) \\ &\quad + \frac{1}{1 - t^2}\left[n(n + 2) + \frac{3}{2\,(1 - t^2)}\left(1 - \frac{t^2}{2}\right)\right]uv\,. \end{aligned} \tag{9.181}$$

To derive the second kind of Chebyshev oscillator in the self-adjoint form we find the integration factor as

$$\mu = \left(t^2 - 1\right)^{3/2} \tag{9.182}$$

and the corresponding Lagrangian function

$$L = \frac{\left(t^2 - 1\right)^{3/2}}{2}\left(\dot{u}^2 - \frac{n(n+2)}{1-t^2}u^2\right) . \tag{9.183}$$

From the generalized momentum,

$$p_u = \left(t^2 - 1\right)^{3/2}\dot{u} , \tag{9.184}$$

we get the self-adjoint Hamiltonian of the second kind of Chebyshev damped oscillator,

$$H(u, p_u) = \frac{\left(t^2 - 1\right)^{3/2}}{2}\left(\frac{{p_u}^2}{\left(t^2 - 1\right)^3} + \frac{n(n+2)}{1-t^2}u^2\right) . \tag{9.185}$$

9.2.13 Shifted Chebyshev I

The parameters of the differential operator for the first kind of shifted Chebyshev function are $p(t) = [t(1-t)]^{\frac{1}{2}}$, $q(t) = 0$, $\lambda = n^2$, $w = [t(1-t)]^{-\frac{1}{2}}$. Hence the differential equation is

$$[t(1-t)]^{\frac{1}{2}}\,\ddot{u} + \frac{1}{2}\frac{1-2t}{\sqrt{t(1-t)}}\dot{u} + [t(1-t)]^{-\frac{1}{2}} n^2 u = 0 . \tag{9.186}$$

After rearranging this equation by multiplying it with $[t(1-t)]^{-\frac{1}{2}}$, the oscillatory form of the shifted Chebyshev function is obtained:

$$\ddot{u} + \frac{1-2t}{2t(1-t)}\dot{u} + \frac{n^2}{t(1-t)}u = 0 , \tag{9.187}$$

in which the time dependent damping and the frequency functions are

$$\Gamma \equiv \frac{1}{2}\frac{1-2t}{t(1-t)}, \quad \omega^2 \equiv \frac{n^2}{t(1-t)} .$$

The doubled form of this oscillator is the system

$$\begin{aligned} \ddot{u} + \frac{1-2t}{2t(1-t)}\dot{u} + \frac{n^2}{t(1-t)}u &= 0 , \\ \ddot{v} - \frac{1-2t}{2t(1-t)}\dot{v} + \frac{1}{(1-t)}\left(\frac{n^2}{t} + \frac{1}{2(1-t)}\right)v &= 0 , \end{aligned} \tag{9.188}$$

with the Lagrangian function

$$L = \dot{u}\dot{v} - \frac{1-2t}{4t(1-t)}(v\dot{u} - \dot{v}u) - \frac{1}{(1-t)}\left(\frac{n^2}{t} + \frac{1}{4(1-t)}\right)uv . \tag{9.189}$$

By the momenta in the dual representation, given by

$$p_u = \dot{v} - \frac{1-2t}{4t(1-t)}v\,, \quad p_v = \dot{u} + \frac{1-2t}{4t(1-t)}u\,, \tag{9.190}$$

we derive the Hamiltonian function,

$$\begin{aligned} H(u, v, p_u, p_v) = p_u p_v + \frac{1-2t}{4t(1-t)}(v p_v - u p_u) + \\ + \frac{1}{(1-t)}\left[\frac{n^2}{t} + \frac{1}{2(1-t)}\left(1 + \frac{1-2t}{8}\right)\right] uv\,. \end{aligned} \tag{9.191}$$

The oscillator equation can be transformed into the self-adjoint form using the integration factor

$$\mu = \sqrt{t(t-1)}\,. \tag{9.192}$$

The corresponding Lagrangian,

$$L = \frac{\sqrt{t(t-1)}}{2}\left(\dot{u}^2 - \frac{n^2}{t(1-t)}u^2\right), \tag{9.193}$$

determines the momentum,

$$p_u = \sqrt{t(t-1)}\dot{u} \tag{9.194}$$

and this gives the Hamiltonian in the self-adjoint form

$$H(u, p_u) = \frac{\sqrt{t(t-1)}}{2}\left(\frac{{p_u}^2}{t(1-t)} + \frac{n^2}{t(1-t)}u^2\right). \tag{9.195}$$

10 Riccati representation of time dependent damped oscillators

In the first part of this book we discussed cosmological models where the size of the universe satisfied the Riccati equation, which was linearized in terms of the Schrödinger problem. In the second part we studied general second-order linear differential equations in oscillator representation and found several solutions in terms of special functions. In the present chapter we will construct a nonlinear representation of damped oscillatory models in the form of the Riccati equation and the corresponding exact solutions.

Theorem 10.1. *The damped harmonic oscillator with time dependent damping $\Gamma(t)$ and frequency $\omega^2(t)$, defined as*

$$\ddot{q} + \Gamma(t)\dot{q} + \omega^2(t)q = 0 \tag{10.1}$$

by the substitution

$$\eta(t) = \frac{\dot{q}(t)}{q(t)}, \tag{10.2}$$

can be represented as a nonlinear Riccati equation:

$$\dot{\eta} + \eta^2 + \Gamma(t)\eta + \omega^2(t) = 0\,. \tag{10.3}$$

Proof. To construct the Riccati equation from the time dependent damped harmonic oscillator equation, we can change the variable to $q(t) = e^{\ln q(t)}$ in equation (11.10). The first and the second derivatives of $q = e^{\ln q}$ are given by following equations:

$$\begin{aligned} \frac{dq}{dt} &= (\ln q)_t e^{\ln q}\,, \\ \frac{d^2q}{dt^2} &= \left\{(\ln q)_{tt} + \left[(\ln q)_t\right]^2\right\} e^{\ln q}\,. \end{aligned} \tag{10.4}$$

After substituting into equation (11.10), we obtain

$$\left\{(\ln q)_{tt} + \left[(\ln q)_t\right]^2 + \Gamma(t)(\ln q)_t + \omega^2(t)\right\} e^{\ln q} = 0\,. \tag{10.5}$$

In terms of new function $\eta \equiv \dot{q}/q = (\ln q)_t$, we then get the Riccati equation

$$\dot{\eta} + \eta^2 + \Gamma(t)\eta + \omega^2(t) = 0\,. \qquad \square$$

As a consequence, equation (10.3) is an explicitly solvable nonlinear system for which the time dependent damping and frequency terms are given in Table 9.2. In every particular case described by special functions we have solutions for the linear oscillator equation $q(t)$ and corresponding solutions $\dot{q}(t)/q(t) \equiv \eta(t)$ for the nonlinear Riccati equation. For example, the solution of the Hermite damped oscillator equation, represented by the Hermite polynomial, gives an exact solution of the Hermite-type Riccati equation. In addition, every zero of $q(t)$ determines the pole singularity of $\eta(t)$.

https://doi.org/10.1515/9783110515367-010

Definition 10.1. A point z_0 on a complex plane is called a zero of order m for the complex function $f(z)$ if $f(z)$ is analytic at z_0 and its first $m-1$ derivatives vanish at z_0, but $f^{(m)}(z_0) \neq 0$.

In explicit form we have

$$f(z_0) = f'(z_0) = f''(z_0) = \cdots = f^{(m-1)}(z_0) = 0 \,,$$
$$f^{(m)}(z_0) \neq 0 \,.$$

The Taylor series for $f(z)$ around z_0 then takes the form

$$f(z) = a_m(z - z_0)^m + a_{m+1}(z - z_0)^{m+1} + a_{m+2}(z - z_0)^{m+2} + \ldots$$

or

$$f(z) = (z - z_0)^m \left[a_m + a_{m+1}(z - z_0)^{m+1} + a_{m+2}(z - z_0)^{m+2} + \ldots\right] , \tag{10.6}$$

where $a_m = f^{(m)}(z_0)/m! \neq 0$.

From this representation it is easy to conclude that if complex function $f(z)$ is analytic at z_0, then $f(z)$ has a zero of order m at z_0 if and only if it can be written as

$$f(z) = (z - z_0)^m g(z) \,,$$

where $g(z)$ is analytic at z_0 and $g(z_0) \neq 0$.

In a similar way we can describe the poles of function $f(z)$. A complex function $f(z)$ has a pole of order m at z_0 if and only if in some punctured neighborhood of z_0

$$f(z) = \frac{g(z)}{(z - z_0)^m} \,,$$

where $g(z)$ is analytic at z_0 and $g(z_0) \neq 0$. As is evident, if a complex function $f(z)$ has a zero of order m at z_0, then $1/f(z)$ has a pole of order m at z_0. These results allow us to formulate the next proposition.

Proposition 10.1. *For every zero t_0 of $q(t)$ as a solution of the linear damped oscillator equation (11.10), there is a corresponding pole t_0 of $\eta(t)$ as a solution of the nonlinear Riccati equation (10.3).*

As an example, let $t = t_0$ be a simple zero of an analytic function $q(t)$, so that $q(t) = (t - t_0)f(t)$, where $f(t_0) \neq 0$. Then,

$$\eta(t) = \frac{\dot{q}(t)}{q(t)} = \frac{f(t) + (t - t_0)\dot{f}(t)}{(t - t_0)f(t)} = \frac{g(t)}{(t - t_0)} \,. \tag{10.7}$$

Here we like to stress that if the damped oscillator equation (11.10) is represented in the self-adjoint form for function $Q(t) = \mu(t)q(t)$, where $\mu(t)$ is the integration factor satisfying equation $\dot{\mu} = \Gamma(t)\mu$, then the corresponding Riccati equation for $u(t) =$

$\dot{Q}(t)/Q(t) = \eta(t) + \Gamma(t)$ takes the form

$$\dot{u} + u^2 - \Gamma(t)u + \omega^2(t) - \dot{\Gamma}(t) = 0\,.$$

In this chapter, we are going to describe different types of Riccati equation, corresponding to damped oscillator equations and solutions in term of the special functions.

10.1 Hypergeometric Riccati equation

The damped hypergeometric oscillator is defined as

$$\ddot{q} + \frac{[c-(a+b+1)\,t]}{t\,(1-t)}\,\dot{q} - \frac{ab}{t\,(1-t)}\,q = 0\,, \tag{10.8}$$

and the corresponding hypergeometric Riccati equation is

$$\dot{\eta} + \eta^2 + \frac{[c-(a+b+1)\,t]}{t\,(1-t)}\eta - \frac{ab}{t\,(1-t)} = 0\,. \tag{10.9}$$

The solution of the second-order differential equation (10.8) is given by the hypergeometric function in the power series form:

$$q(t) = {}_2F_1(a,b,c;t) = \sum_{n=0}^{\infty} \frac{(a)_n(b)_n}{(c)_n n!}\,t^n\,. \tag{10.10}$$

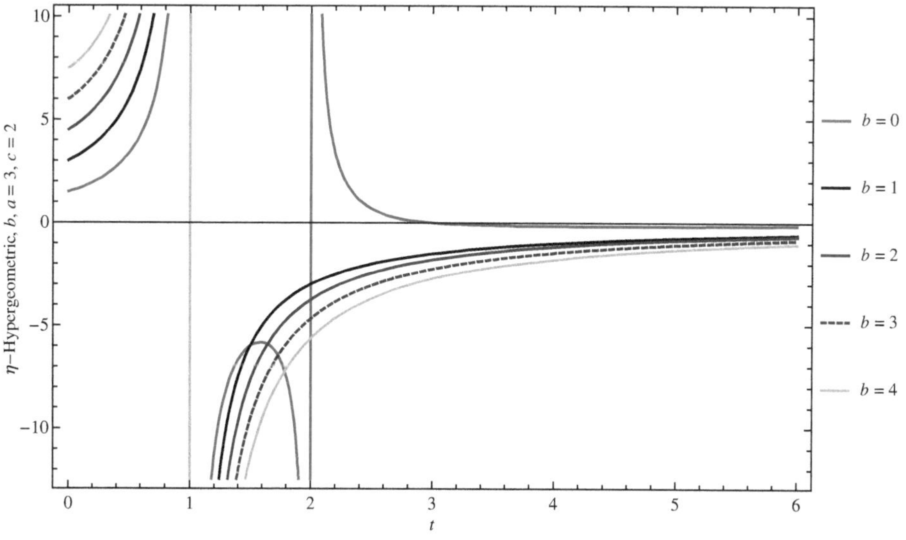

Fig. 10.1: Evolution of poles $\eta(t)$ of the hypergeometric function with coefficients $a = 3$, $c = 2$ and $b = 0, 1, 2, 3, 4$.

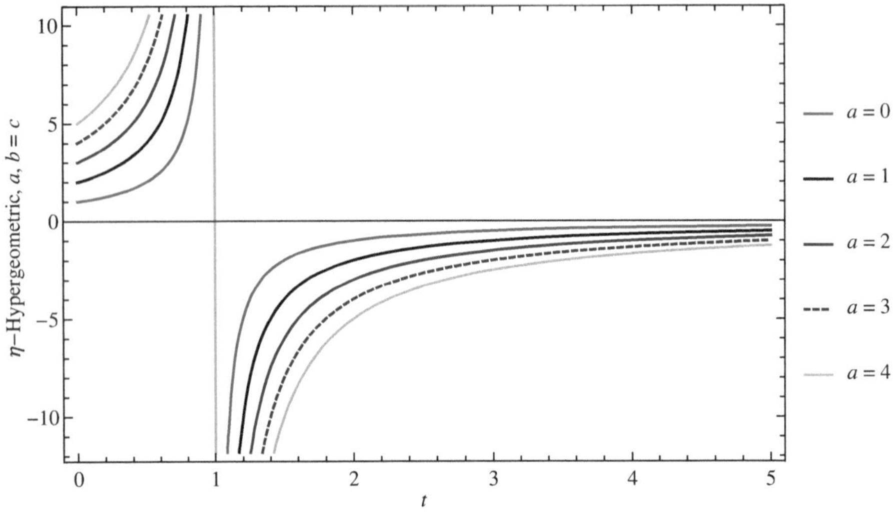

Fig. 10.2: Evolution of simple pole solution $\eta(t)$ of the hypergeometric function with coefficients $b = c$ for $a = 1, 2, 3, 4, 5$.

The corresponding solution of the Riccati equation (10.9) is then

$$\eta(t) = \frac{d}{dt} \ln \, {}_2F_1(a, b, c; t) \, . \tag{10.11}$$

Figure 10.1 shows the evolution of poles of the Hypergeometric Riccati type equation based on equation (10.11) with respect to varying b, constant a and c parameters. According to this, at time $t = 1$, the hyperbolic Riccation equation results in hyperbolic solutions. If we consider the particular case of this solution by choosing in ${}_2F_1(a, b, c; t)$ parameters $b = c$, then the solution of the oscillator equation becomes

$$q(t) =_2 F_1(a, b, b, t) = \frac{1}{(1-t)^a} \, , \tag{10.12}$$

and for the Riccati equation we get the following simple solution:

$$\eta(t) = \frac{a}{1-t} \, . \tag{10.13}$$

Figure 10.2 depicts the solution of the Hypergeometric Riccati type equation in the case of $b = c$ with varying a parameter. In this case, the evolution of poles only dependents on time and parameter a (as in equation (10.13)). As is seen in Figure 10.2, the singularity again emerges at $t = 1$.

10.2 Confluent hypergeometric Riccati equation

The damped confluent hypergeometric oscillator is written as

$$\ddot{q} + \frac{(b-t)}{t}\dot{q} - \frac{a}{t}q = 0\,,$$

and the corresponding confluent hypergeometric Riccati equation is

$$\dot{\eta} + \eta^2 + \frac{(b-t)}{t}\eta - \frac{a}{t} = 0\,. \tag{10.14}$$

Many special functions of mathematical physics can be represented in terms of hypergeometric or confluent hypergeometric functions. In the following subsections we present Riccati equations corresponding to damped oscillators of the special functions and discuss relations between the zeros and poles of the oscillator-type and the Riccati-type solutions. In all examples given here, the zeros are related to the degree of the polynomial solution, and in the Riccati representation we have poles related to the degree of these polynomials.

10.3 Legendre-type Riccati equation

The damped Legendre oscillator equation is defined as

$$\ddot{q} - \frac{2t}{1-t^2}\dot{q} + \frac{n(n+1)}{1-t^2}q = 0\,,$$

which determines the Legendre-type Riccati equation

$$\dot{\eta} + \eta^2 - \frac{2t}{1-t^2}\eta + \frac{n(n+1)}{1-t^2} = 0\,. \tag{10.15}$$

For the Legendre polynomials $P_n(t)$, sometimes called Legendre functions of the first kind, the recurrence relation is given by

$$\dot{P}_n(t) = \frac{nt}{t^2-1}P_n - \frac{n}{t^2-1}P_{n-1}\,, \quad n = 1, 2, \ldots \tag{10.16}$$

Therefore, for the corresponding solution of the Riccati equation we have

$$\eta(t) = \frac{\dot{q}(t)}{q(t)} = \frac{\dot{P}_n(t)}{P_n(t)} = \frac{nt}{t^2-1}\left[t - \frac{P_{n-1}(t)}{P_n(t)}\right]\,. \tag{10.17}$$

For the first few Legendre polynomials

$$
\begin{aligned}
P_0 &= 1 \\
P_1 &= t \\
P_2 &= \frac{1}{2}\left(3t^2 - 1\right) && (10.18) \\
P_3 &= \frac{1}{2}\left(5t^3 - 3t\right) \\
P_4 &= \frac{1}{8}\left(35t^4 - 30t^2 + 3\right) && (10.19) \\
&\dots
\end{aligned}
$$

we get several solutions, starting with

$$P_1(t) = t \quad \rightarrow \quad \eta(t) = \frac{\dot{P}_1(t)}{P_1(t)} = \frac{1}{t}\,, \tag{10.20}$$

where $t = 0$ is a simple zero of $P_1(t)$ and a simple pole of $\eta(t)$. The next solution is

$$P_2(t) = \frac{1}{2}\left(3t^2 - 1\right) \quad \rightarrow \quad \eta(t) = \frac{\dot{P}_2(t)}{P_2(t)} = \frac{6t}{3t^2 - 1}\,, \tag{10.21}$$

where $t = \pm\sqrt{\frac{1}{3}}$ are two simple zeros of $P_2(t)$ and two simple poles of $\eta(t)$. Continuing with these examples, from (10.17) we can see that n zeros of the n-th Legendre polynomial $P_n(t)$ correspond to the n poles of the Legendre-type Riccati equation.

Figure 10.3 presents the evolution of the η functions which are the solutions of the Legendre type Riccati equation in terms of different Legendre polynomials given

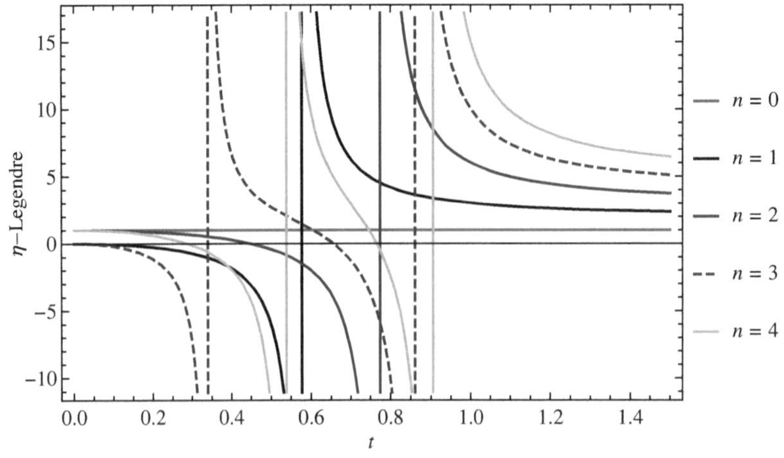

Fig. 10.3: Evolution of the pole solutions of the Legendre polynomials for $n = 0, 1, 2, 3$, and 4 (based on equation (10.17).

by $n = 0, 1, 2, 3, 4$. As is seen in Figure 10.3, while the Legendre polynomial with $n = 0$ shows contant change in time, values $n = 1, 2$ indicate standard hyperbolic function in time. On the other hand, when n accepts values like 3 and 4, the number of the singularities that form hyperbolic functions increases.

10.4 Associated Legendre-type Riccati equation

To the associated Legendre oscillator equation

$$\ddot{q} - \frac{2t}{1-t^2}\dot{q} + \frac{1}{1-t^2}\left[l(l+1) - \frac{m^2}{1-t^2}\right] q = 0\,,$$

corresponds the following Riccati equation:

$$\dot{\eta} + \eta^2 - \frac{2t}{1-t^2}\eta + \frac{1}{1-t^2}\left[l(l+1) - \frac{m^2}{1-t^2}\right] = 0\,. \tag{10.22}$$

The associated Legendre polynomials $P_l{}^m(t)$ satisfy the recurrence relation

$$\dot{P}_l{}^m(t) = -\frac{1}{\sqrt{1-t^2}}P_l{}^{m+1}(t) + \frac{mt}{t^2-1}P_l{}^m(t)\,, \tag{10.23}$$

where $-l \leq m \leq l$. Hence, we obtain solutions of the associated Legendre-type Riccati equation as

$$\eta(t) = \frac{\dot{q}(t)}{q(t)} = \frac{\dot{P}_l^m(t)}{P_l{}^m(t)} = \frac{mt}{t^2-1} - \frac{1}{\sqrt{1-t^2}}\frac{P_l{}^{m+1}}{P_l{}^m}\,. \tag{10.24}$$

Figure 10.4 shows four panels. In each panel it is presented how the solutions of the Associated Legendre type Riccati equation evolves in time with varying l values ($n = 0, 1, 2$), and fixed m index.

10.5 Hermite-type Riccati equation

The Hermite oscillator equation

$$\ddot{q} - 2t\dot{q} + \lambda q = 0$$

in the Riccati representation is written as

$$\dot{\eta} + \eta^2 - 2t\eta + \lambda = 0\,. \tag{10.25}$$

The recurrence relation for the Hermite polynomials,

$$\dot{H}_n(t) = 2nH_{n-1}\,, \tag{10.26}$$

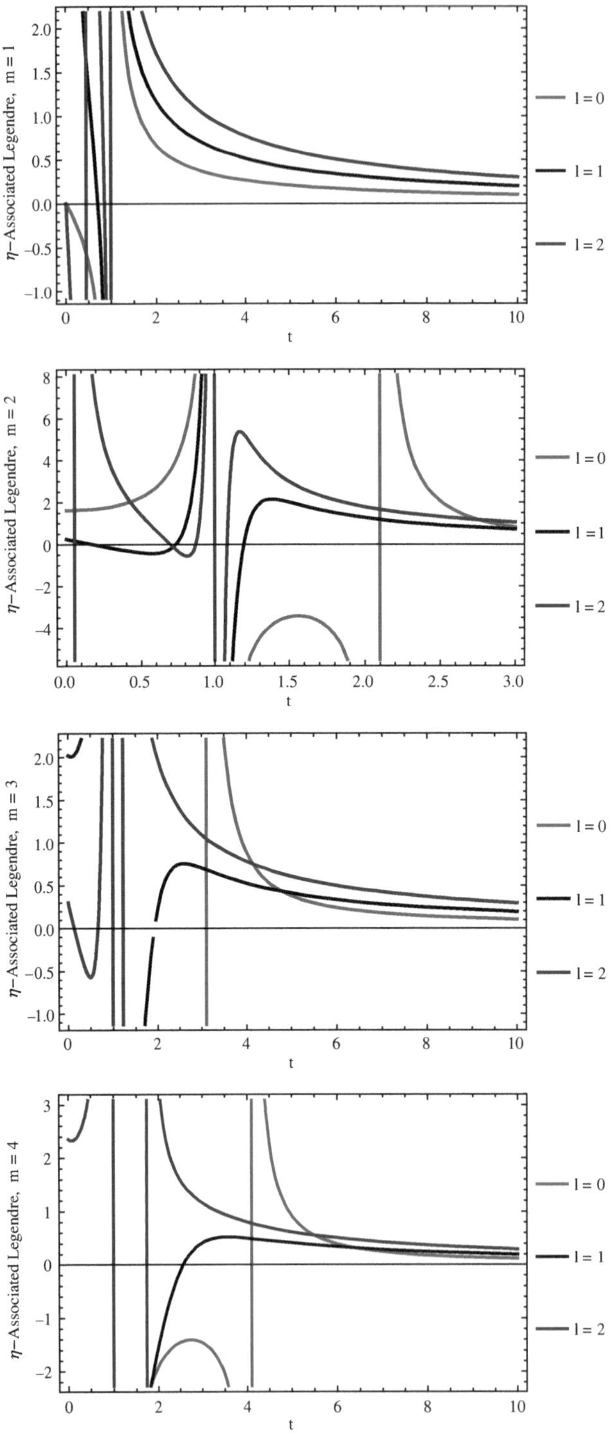

Fig. 10.4: Evolution of the pole solutions of the associated Legendre polynomials at indices $n = 4$ and for each $m = 1, 2, 3, 4$ from left to right and up to down.

then gives us the following solution of the Hermite-type Riccati equation:

$$\eta(t) = \frac{\dot{q}(t)}{q(t)} = \frac{\dot{H}_n(t)}{H_n(t)} = 2n\frac{H_{n-1}(t)}{H_n(t)} . \tag{10.27}$$

Figure 10.5 presents the Riccati type solution of the Hermite oscillator in terms of five differemt n values. As is seen in Figure 10.5, the number of oscillations increases with increasing n index between $t = 0$ and $t = 2$.

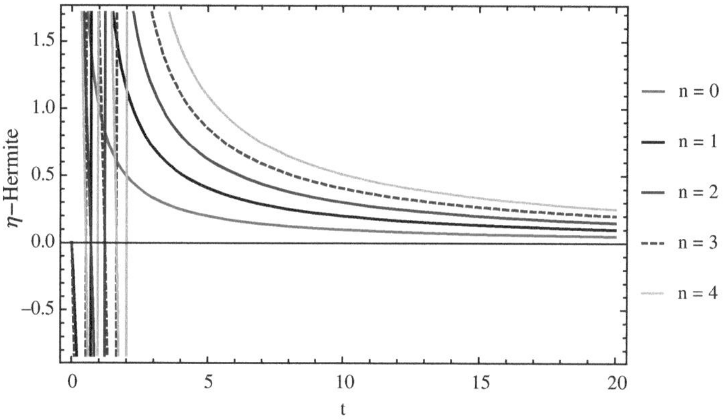

Fig. 10.5: Evolution of the pole solutions $\eta(t)$ of the Hermite polynomials for $n = 0, 1, 2, 3,$ and 4 (based on equation (10.27)).

10.6 Laguerre-type Riccati equation

The Laguerre damped oscillator equation is written as

$$\ddot{q} + \frac{(1-t)}{t}\dot{q} + \frac{\alpha}{t}, q = 0 ,$$

where α is a nonnegative constant. The Riccati representation of this equation then becomes

$$\dot{\eta} + \eta^2 + \frac{(1-t)}{t}\eta + \frac{\alpha}{t} = 0 . \tag{10.28}$$

The recurrence relation for the Laguerre polynomials

$$\dot{L}_\alpha(t) = \frac{\alpha}{t}\left(L_\alpha(t) - L_{\alpha-1}(t)\right) , \tag{10.29}$$

gives us the following solutions of the Riccati equation:

$$\eta(t) = \frac{\dot{q}(t)}{q(t)} = \frac{\dot{L}_\alpha(t)}{L_\alpha(t)} = \frac{\alpha}{t}\left[1 - \frac{L_{\alpha-1}(t)}{L_\alpha(t)}\right] . \tag{10.30}$$

Figure 10.6 shows the dynamical behaviour the Riccati type solution of the Laguerre equation in terms of five different n values. Figure 10.6 represents two different type of singularity for each Laguerre polynomial; the first singulalrity occurs at $t = 0$ while second type of singularity occurs due to the degree of the polynomial.

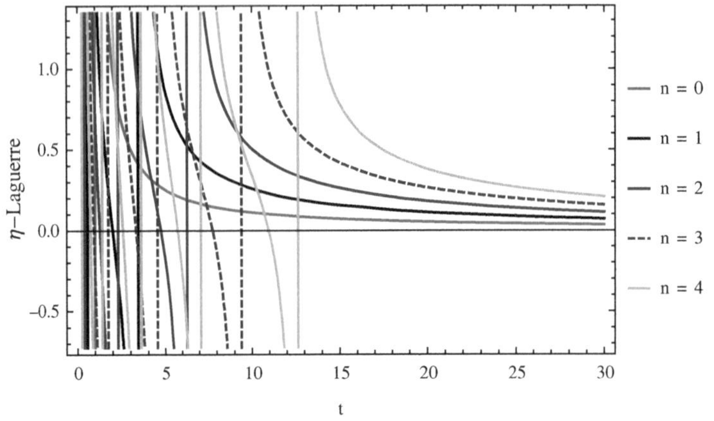

Fig. 10.6: Evolution of the pole solution $\eta(t)$ of the Laguerre polynomials for $n = 0, 1, 2, 3$, and 4 (based on equation (10.30)).

10.7 Associated Laguerre-type Riccati equation

The associated Laguerre oscillator equation is written in the following form:

$$\ddot{q} + \frac{k+1-t}{t}\dot{q} + \frac{n}{t}q = 0 .$$

The Riccati representation of this equation is

$$\dot{\eta} + \eta^2 + \frac{k+1-t}{t}\eta + \frac{n}{t} = 0 . \tag{10.31}$$

The associated Laguerre polynomials $L_n^k(t)$ satisfy the recurrence relations

$$\dot{L_n^k}(t) = \frac{n}{t}L_n^k(t) - \frac{(n+k)}{t}L_{n-1}^k(t) , \tag{10.32}$$

giving solutions of the Riccati equation in the form

$$\eta(t) = \frac{\dot{q}(t)}{q(t)} = \frac{1}{t}\left[n - (n+k)\frac{L_{n-1}^k(t)}{L_n^k(t)}\right] . \tag{10.33}$$

The panels of Figure 10.7 show the evolution of the associated Laguerre oscillator in terms of Riccati solution. As is seen, parameter k increases in panels from left to right,

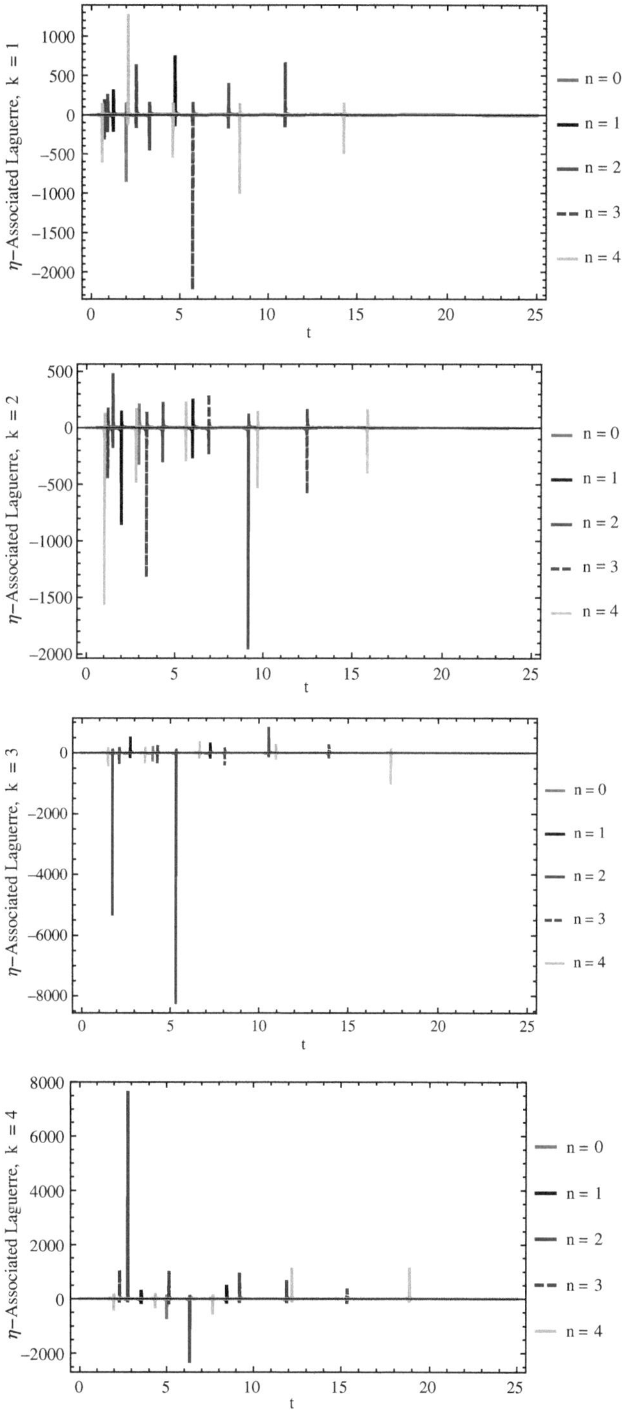

Fig. 10.7: Evolution of the pole solutions of the associated Laguerre polynomials at indices $n = 0, \dots, 4$ and for each $m = 1, 2, 3, 4$ from left to right and up to down.

and top to bottom with varying n index. According to this, the amplitude of the oscillations of each polynomial, the position of peaks change in terms of k index.

10.8 Chebyshev I-type Riccati equation

The oscillator representation of the first kind of Chebyshev equation is given by

$$\ddot{q} - \frac{t}{1-t^2}\dot{q} + \frac{n^2}{1-t^2}q = 0\,,$$

and its Riccati representation is

$$\dot{\eta} + \eta^2 - \frac{t}{1-t^2}\eta + \frac{n^2}{1-t^2} = 0. \tag{10.34}$$

The recurrence relation for the first kind of Chebyshev polynomials is

$$\dot{T}_n(t) = \frac{n}{1-t^2}\left[T_{n-1}(t) - tT_n(t)\right] \tag{10.35}$$

and the corresponding solutions of the Riccati equation are

$$\eta(t) = \frac{\dot{q}(t)}{q(t)} = \frac{\dot{T}_n(t)}{T_n(t)} = \frac{n}{1-t^2}\left[\frac{T_{n-1}(t)}{T_n(t)} - t\right]. \tag{10.36}$$

The dynamical behaviour of the poles of the first kind of Chebyshev oscillator with different polynomial index n is shown in Figure 10.8.

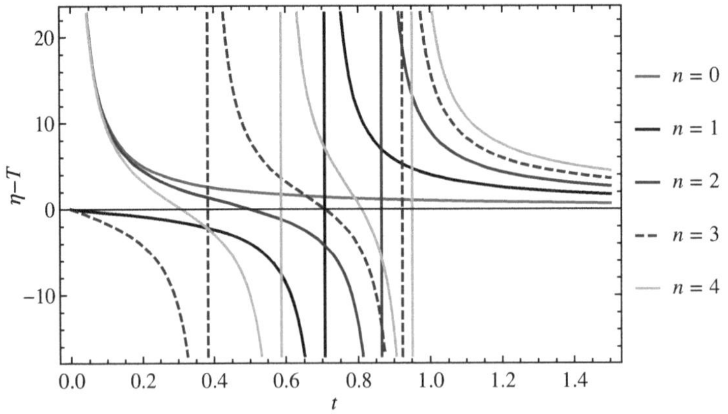

Fig. 10.8: Evolution of the pole solution $\eta(t)$ of the first kind of Chebyshev polynomials for $n = 0, 1, 2, 3$, and 4 (based on equation (10.36)).

10.9 Chebyshev II-type Riccati equation

The oscillator representation of the second kind of Chebyshev equation,

$$\ddot{q} - \frac{3t}{1-t^2}\dot{q} + \frac{n(n+2)}{1-t^2}q = 0\,,$$

gives the Riccati representation

$$\dot{\eta} + \eta^2 - \frac{3t}{1-t^2}\eta + \frac{n(n+2)}{1-t^2} = 0\,. \tag{10.37}$$

The recurrence relation for polynomial solutions

$$\dot{U}_n(t) = \frac{n+1}{1-t^2}U_{n-1} - \frac{nt}{1-t^2}U_n(t) \tag{10.38}$$

gives the following solutions of the Riccati equation:

$$\eta(t) = \frac{\dot{q}(t)}{q(t)} = \frac{\dot{U}_n(t)}{U_n(t)} = -\frac{n}{1-t^2}\left[t - \frac{U_{n-1}(t)}{U_n(t)}\right]\,. \tag{10.39}$$

Figure 10.9 presents the evolution of Riccati type solution of the second kind Chebyshev oscillator in which poles directly given by the number of zeros of the second kind of Chebyshev polynomials ($n = 0, 1, 2\ldots$).

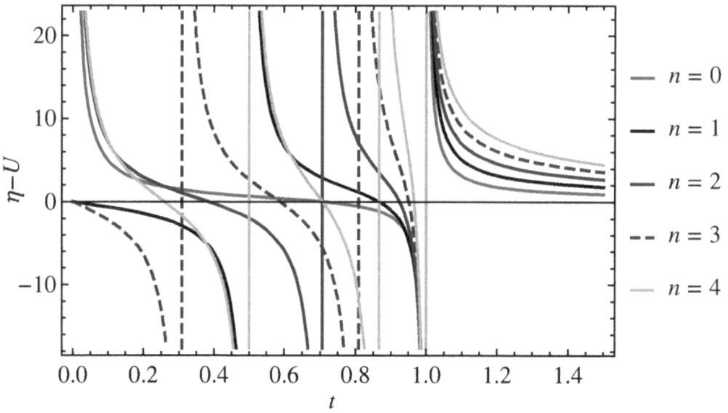

Fig. 10.9: Evolution of the pole solutions $\eta(t)$ of the second kind of Chebyshev polynomials for $n = 0$, 1, 2, 3, and 4 (based on equation (10.39)).

11 Quantization of the harmonic oscillator with time dependent parameters

The harmonic oscillator with time dependent frequency and damping has recently attracted considerable interest. Apart from its intrinsic mathematical interest, this problem has attracted much attention because of its connection with many other problems belonging to different areas of physics, such as plasma physics, gravitation, quantum optics, etc. For example, [29] studied a harmonic oscillator with a time dependent frequency and a constant mass in an expanding universe.

In this section we are mainly concerned with the damped harmonic oscillator with both frequency and damping being arbitrary functions of time. Our main purpose is to exhibit in a simple way an alternative treatment for such a system. The treatment is based on the use of some ansatz for the wave function, after [24], and dynamical symmetry of the Schrödinger equation. The solution of the quantum problem is thus reduced to solving the classical damped oscillator and the Riccati equation.

We are going to quantize the Hamiltonian, which is produced from the self-adjoint operator, and will work in the Schrödinger picture. Firstly, the classical Hamiltonian function of the quantum damped harmonic oscillator is given as

$$H = \frac{p^2}{2\mu(t)} + \omega^2(t)\frac{\mu(t)}{2}q^2 . \tag{11.1}$$

In the canonical quantization approach to this system we define the coordinate x and the momentum p operators as

$$p \to -i\hbar\frac{\partial}{\partial x} , \quad q \to x . \tag{11.2}$$

The operators p and q satisfy the canonical commutation relations

$$[p, x] = -i\hbar , \quad [x, x] = 0 = [p, p] . \tag{11.3}$$

Evolution of the wave function $\Psi(x, t)$ is described by the time dependent Schrödinger equation,

$$i\hbar\frac{\partial\Psi(x, t)}{\partial t} = H\Psi(x, t) ,$$

with the above Hamiltonian operator (11.1),

$$i\hbar\frac{\partial\Psi(x, t)}{\partial t} = -\frac{\hbar^2}{2\mu(t)}\frac{\partial^2\Psi(x, t)}{\partial x^2} + \omega^2(t)\frac{\mu(t)}{2}x^2\Psi(x, t) . \tag{11.4}$$

https://doi.org/10.1515/9783110515367-011

11.1 Gaussian wave function

We search for a particular solution of the Schrödinger equation (11.4) in the form of the time dependent Gaussian wave packet,

$$\Psi(x,t) = \exp\frac{i}{2\hbar}\left[a(t)x^2 + 2b(t)x + c(t)\right] , \tag{11.5}$$

where $a(t)$, $b(t)$ and $c(t)$ are unknown functions to be determined. Substituting this wave function into (11.4) we get the system of ordinary differential equations for these functions:

$$\dot{a}(t) + \frac{a^2(t)}{\mu(t)} + \omega^2(t)\mu(t) = 0 , \tag{11.6}$$

$$\dot{b}(t) + \frac{a(t)}{\mu(t)}b(t) = 0 , \tag{11.7}$$

$$\dot{c} - i\hbar\frac{a(t)}{\mu(t)} + \frac{1}{\mu(t)}b^2(t) = 0 . \tag{11.8}$$

The first equation in this system is the nonlinear Riccati equation. It can be linearized by the following substitution:

$$a(t) = \mu(t)\frac{\dot{q}(t)}{q(t)} . \tag{11.9}$$

Theorem 11.1. *Every solution of the linear damped oscillator with time dependent damping $\Gamma(t)$ and frequency $\omega^2(t)$, defined as*

$$\ddot{q} + \Gamma(t)\dot{q} + \omega^2(t)q = 0 , \tag{11.10}$$

where

$$\Gamma(t) = \frac{\dot{\mu}(t)}{\mu(t)} , \tag{11.11}$$

gives the following solution of the nonlinear Riccati equation:

$$\dot{a}(t) + \frac{a^2(t)}{\mu(t)} + \omega^2(t)\mu(t) = 0 , \tag{11.12}$$

where $a(t)$ is given by

$$a(t) = \mu(t)\frac{\dot{q}(t)}{q(t)} . \tag{11.13}$$

Proof. By direct substitution of (11.13) into the Riccati equation we find that $q(t)$ satisfies the linear equation

$$\ddot{q} + \frac{\dot{\mu}(t)}{\mu(t)}\dot{q} + \omega^2(t)q = 0$$

of the damped oscillator. □

Using this theorem we can use the special functions damped oscillator equations, developed in Chapters 9 and 10, to construct corresponding solutions of the Riccati equation (11.6). Then, the linear equations (11.7) and (11.8), for functions $b(t)$ and $c(t)$ respectively, can be easily solved. By direct integration we get the following theorem:

Theorem 11.2. *The solution of the system (11.7) and (11.8) in terms of $q(t)$, with initial values $q(t_0)$, $b(t_0)$ and $c(t_0)$ at time t_0, is*

$$b(t) = b(t_0)\frac{q(t_0)}{q(t)}, \tag{11.14}$$

$$c(t) = c(t_0) + i\hbar \ln\frac{q(t)}{q(t_0)} - b^2(t_0)q^2(t_0)\int_{t_0}^{t}\frac{d\xi}{\mu(\xi)q^2(\xi)}. \tag{11.15}$$

Combining these results together we finally get the next theorem.

Theorem 11.3. *The Gaussian wave packet solution,*

$$\Psi(x,t) = \exp\frac{i}{2\hbar}\left[a(t)x^2 + 2b(t)x + c(t)\right], \tag{11.16}$$

of the time dependent Schrödinger equation for the quantum damped oscillator,

$$i\hbar\frac{\partial\Psi(x,t)}{\partial t} = -\frac{\hbar^2}{2\mu(t)}\frac{\partial^2\Psi(x,t)}{\partial x^2} + \omega^2(t)\frac{\mu(t)}{2}x^2\Psi(x,t), \tag{11.17}$$

is completely determined by a solution of the classical damped oscillator equation,

$$\ddot{q} + \frac{\dot{\mu}(t)}{\mu(t)}\dot{q} + \omega^2(t)q = 0, \tag{11.18}$$

according to formulas

$$a(t) = \mu(t)\frac{\dot{q}(t)}{q(t)}, \tag{11.19}$$

$$b(t) = b(t_0)\frac{q(t_0)}{q(t)}, \tag{11.20}$$

$$c(t) = c(t_0) + i\hbar \ln\frac{q(t)}{q(t_0)} - b^2(t_0)q^2(t_0)\int_{t_0}^{t}\frac{d\xi}{\mu(\xi)q^2(\xi)}. \tag{11.21}$$

11.2 Dynamical symmetry and exact solutions

In previous section we constructed a solution of the Schrödinger equation in the Gaussian form, which has no zeros. Here, to obtain other exact solutions we apply the method of dynamical symmetry [32]. We consider the Schrödinger equation (11.17) in the operator form,

$$S\Psi(x,t) = 0, , \tag{11.22}$$

where the Schrödinger operator for the damped oscillator is

$$S \equiv i\hbar\frac{\partial}{\partial t} + \frac{\hbar^2}{2\mu(t)}\frac{\partial^2}{\partial x^2} - \omega^2(t)\frac{\mu(t)}{2}x^2 \,. \tag{11.23}$$

Theorem 11.4. *If an operator K is commutative with the Schrödinger operator S,*

$$[K, S] = K\,S - S\,K = 0 \,, \tag{11.24}$$

and $\Psi(x,t)$ is a solution of the Schrödinger equation,

$$S\Psi(x,t) = 0 \,, \tag{11.25}$$

then function

$$\Phi(x,t) = K\Psi(x,t) \tag{11.26}$$

is also solution of the Schrödinger equation,

$$S\Phi(x,t) = 0 \,. \tag{11.27}$$

Proof. Applying the commutator to the state $\Psi(x,t)$ we have desired result:

$$0 = K\,S\Psi(x,t) = S\,K\Psi(x,t) = S\,\Phi(x,t) = 0 \,. \tag{11.28}$$

□

Definition 11.1. Operator K is called the dynamical symmetry operator for the Schrödinger equation.

If we know operator K, then we can construct solutions of the Schrödinger equation by applying this operator several times to a known solution of the equation.

Theorem 11.5. *The generalized boost operator,*

$$K = A(t)x + iB(t)\frac{d}{dx} \,, \tag{11.29}$$

is the dynamical symmetry operator for the Schrödinger equation (11.17), if functions $A(t)$ and $B(t)$ are solutions of the first-order system

$$\hbar\dot{A}(t) + \omega^2(t)\mu(t)B(t) = 0 \,, \tag{11.30}$$

$$\mu(t)\dot{B}(t) - \hbar A(t) = 0 \,. \tag{11.31}$$

Proof. By direct computation of the commutator $[S, K]$, and using

$$\left[\frac{d^2}{dx^2}, x\right] = 2\frac{d}{dx}$$

and

$$\left[x^2, \frac{d}{dx}\right] = -2x \,,$$

we find the system (11.30), (11.31). Differentiating this system in time we find that functions $A(t)$ and $B(t)$ satisfy the damped oscillator equations, respectively

$$\ddot{A} - \left(\frac{\dot{\mu}(t)}{\mu(t)} + 2\frac{\dot{\omega}}{\omega}\right)\dot{A} + \omega^2(t)A = 0\,, \tag{11.32}$$

$$\ddot{B} + \frac{\dot{\mu}(t)}{\mu(t)}\dot{B} + \omega^2(t)B = 0\,. \tag{11.33}$$

□

Theorem 11.6. *Every solution $A(t)$ and $B(t)$ of the damped oscillator equations (11.32), (11.33), satisfying an additional constraint*

$$\mu(t)\dot{B}(t) = \hbar A(t)\,, \tag{11.34}$$

determines the dynamical symmetry operator

$$K = A(t)x + iB(t)\frac{d}{dx}\,. \tag{11.35}$$

Theorem 11.7. *If $\Psi_0(x,t)$ is a solution of the Schrödinger equation (11.17), then every function*

$$\Psi_n(x,t) = K^n\Psi(x,t), \quad n = 1, 2, \ldots \tag{11.36}$$

is also a solution of this equation.

11.3 Examples of exact solutions

Here we illustrate the above results on two examples: harmonic and Caldirola–Kanai damped oscillators.

11.3.1 Harmonic oscillator

As a first example we consider the harmonic oscillator with constant frequency,

$$\ddot{q} + \omega_0^2 q = 0\,, \tag{11.37}$$

which admits the particular solution

$$q(t) = q_0 e^{i\omega_0 t}\,, \tag{11.38}$$

with initial value $q(0) \equiv q_0$. Due to the absence of the damping term, we can choose the integration factor $\mu(t) = 1$. Then, the solution of the Riccati equation is

$$a(t) = \frac{\dot{q}(t)}{q(t)} = i\omega_0\,. \tag{11.39}$$

Corresponding solutions for $b(t)$ and $c(t)$ are

$$b(t) = b_0 e^{-i\omega_0 t} , \tag{11.40}$$

$$c(t) = c_0 - \hbar\omega_0 t - b_0^2 \frac{1 - e^{-2i\omega_0 t}}{2i\omega_0} . \tag{11.41}$$

For simplicity we choose initial values $b_0 = 0$ and $c_0 = 0$, so that

$$b(t) = 0 , \tag{11.42}$$

$$c(t) = -\hbar\omega_0 t . \tag{11.43}$$

For the Gaussian wave packet (11.5) we then have the well-known solution

$$\Psi_0(x,t) = \exp\left(-i\frac{\omega_0}{2}t\right)\exp\left(-\frac{\omega_0}{2\hbar}x^2\right) , \tag{11.44}$$

describing the ground state of the quantum oscillator with energy

$$E_0 = \frac{\hbar\omega_0}{2} . \tag{11.45}$$

To construct higher energy states we need to find the dynamical symmetry operator K, which is determined by solutions of two oscillator equations

$$\ddot{A} + \omega_0^2 A = 0 , \tag{11.46}$$

$$\ddot{B} + \omega_0^2 B = 0 \tag{11.47}$$

constrained by equation (11.34). Hence, we have the solution

$$A(t) = A_0 e^{-i\omega_0 t} , \tag{11.48}$$

$$B(t) = \frac{i\hbar}{\omega_0} A_0 e^{-i\omega_0 t} . \tag{11.49}$$

Combining these solutions we find the boost operator in the following form:

$$K = A_0 e^{-i\omega_0 t}\left(x - \frac{\hbar}{\omega_0}\frac{d}{dx}\right) . \tag{11.50}$$

Applying this operator to the ground state (11.44), we find the wave function,

$$\Psi_1(x,t) = K\Psi_0(x,t) = A_0 e^{-i\frac{3}{2}\omega_0 t}\left(x - \frac{\hbar}{\omega_0}\frac{d}{dx}\right) e^{-\frac{\omega_0}{2\hbar}x^2} , \tag{11.51}$$

for the first energy level:

$$E_1 = \frac{3\hbar\omega_0}{2} .$$

After application of the K operator n times to the ground state wave function, we get the wave function of the quantum harmonic oscillator,

$$\Psi_n(x,t) = K^n\Psi_0(x,t) = A_0 e^{-i\frac{2n+1}{2}\omega_0 t}\left(x - \frac{\hbar}{\omega_0}\frac{d}{dx}\right)^n e^{-\frac{\omega_0}{2\hbar}x^2} , \tag{11.52}$$

for the state with energy

$$E_n = \hbar\omega_0\left(n + \frac{1}{2}\right).$$

The solution can be rewritten, in a very well-known form, in terms of the Hermite polynomials:

$$\Psi_n(x,t) = C_0 e^{-i(n+\frac{1}{2})\omega_0 t} H_n\left(x\sqrt{\frac{\omega_0}{\hbar}}\right) e^{-\frac{\omega_0}{2\hbar}x^2}. \tag{11.53}$$

11.3.2 Caldirola–Kanai damped oscillator

As a second application we consider the damped oscillator with constant damping Γ and frequency ω_0,

$$\ddot{q} + \Gamma\dot{q} + \omega_0^2 q = 0. \tag{11.54}$$

In the self-adjoint form it is the Caldirola–Kanai model with integration factor

$$\mu(t) = e^{\Gamma t}.$$

For the underdamping case, we choose the particular solution of this equation in the form

$$q(t) = q_0 e^{-\frac{\Gamma}{2}t + i\Omega t}, \tag{11.55}$$

where

$$\Omega^2 = \omega_0^2 - \frac{\Gamma^2}{4}.$$

This determines the solution of the Riccati equation in the form

$$a(t) = e^{\Gamma t}\frac{\dot{q}(t)}{q(t)} = e^{\Gamma t}\left(-\frac{\Gamma}{2} + i\Omega\right).$$

For functions $b(t)$ and $c(t)$ with initial values $b_0 = 0$ and $c_0 = 0$ we have

$$b(t) = 0, \tag{11.56}$$

$$c(t) = -\hbar\Omega t - i\hbar\frac{\Gamma}{2}t. \tag{11.57}$$

We then get exact solution of the Schrödinger equation for the ground state as follows:

$$\Psi_0(x,t) = \exp\left(\frac{\Gamma}{4}t - i\frac{\Omega}{2}t\right)\exp\left(-\frac{i\Gamma}{4\hbar}e^{\Gamma t}x^2 - \frac{\Omega}{2\hbar}e^{\Gamma t}x^2\right). \tag{11.58}$$

To construct the wave functions for excited states we need to find the operator K, which is determined by the Bateman dual oscillator system:

$$\ddot{A} - \Gamma\dot{A} + \omega_0^2 A = 0, \tag{11.59}$$

$$\ddot{B} - \Gamma\dot{B} + \omega_0^2 B = 0. \tag{11.60}$$

The proper solution of this equations, satisfying the constraint, is

$$A(t) = A_0 \exp\left(\frac{\Gamma}{2}t - i\Omega t\right), \tag{11.61}$$

$$B(t) = -A_0 \frac{\hbar}{\omega_0^2}\left(\frac{\Gamma}{2} - i\Omega\right) \exp\left(\frac{\Gamma}{2}t - i\Omega t\right). \tag{11.62}$$

This gives us the boost operator in the form

$$K = e^{-i\Omega t}\left(e^{\frac{\Gamma}{2}t}x - \frac{i\hbar}{\omega_0^2}\left(\frac{\Gamma}{2} - i\Omega\right) e^{-\frac{\Gamma}{2}t}\frac{d}{dx}\right). \tag{11.63}$$

Bibliography

[1] E. Alfinito, R. Manka and G. Vitiello. Double Universe. *ArXiv High Energy Physics – Theory e-prints*, May 1997.

[2] E. Alfinito, R. Manka and G. Vitiello. Vacuum structure for expanding geometry. *Classical and Quantum Gravity*, 17:93–111, January 2000.

[3] E. Alfinito and G. Vitiello. Canonical quantization and expanding metrics. *Physics Letters A*, 252:5–10, February 1999.

[4] E. Alfinito and G. Vitiello. Double universe and the arrow of time. In *Journal of Physics Conference Series*, volume 67 of *Journal of Physics Conference Series*, page 012010, May 2007.

[5] H. Bateman. On Dissipative Systems and Related Variational Principles. *Physical Review*, 38:815–819, August 1931.

[6] C. L. Bennett, D. Larson, J. L. Weiland, N. Jarosik, G. Hinshaw, N. Odegard, K. M. Smith, R. S. Hill, B. Gold, M. Halpern, E. Komatsu, M. R. Nolta, L. Page, D. N. Spergel, E. Wollack, J. Dunkley, A. Kogut, M. Limon, S. S. Meyer, G. S. Tucker and E. L. Wright. Nine-year Wilkinson Microwave Anisotropy Probe (WMAP) Observations: Final Maps and Results. *Astrophysical Journal, Supplement*, 208:20, October 2013.

[7] L. Bianchi. On the three-dimensional spaces which admit a continuous group of motions. *Memorie di Matematica e di Fisica della Società Italiana delle Scienze, Serie Terza XI, 267-352 (1898)*, 11:267–352, 1898.

[8] M. Blasone, E. Graziano, O. K. Pashaev and G. Vitiello. Dissipation and Topologically Massive Gauge Theories in the Pseudo-Euclidean Plane. *Annals of Physics*, 252:115–132, November 1996.

[9] R. H. Brandenberger. Quantum fluctuations as the source of classical gravitational perturbations in inflationary universe models. *Nuclear Physics B*, 245:328–342, 1984.

[10] K. A. Bronnikov, E. N. Chudayeva and G. N. Shikin. Magneto-dilatonic Bianchi-I cosmology: isotropization and singularity problems. *Classical and Quantum Gravity*, 21:3389–3403, July 2004.

[11] P. Caldirola. Forze non conservative nella meccanica quantistica. *Il Nuovo Cimento*, 18:393–400, November 1941.

[12] V. Canuto, P. J. Adams, S.-H. Hsieh and E. Tsiang. Scale-covariant theory of gravitation and astrophysical applications. *Physical Review D*, 16:1643–1663, September 1977.

[13] J. C. Carvalho. Unified description of the early universe. *International Journal of Theoretical Physics*, 35:2019–2028, September 1996.

[14] E. Celeghini, M. Rasetti and G. Vitiello. Squeezing and quantum groups. *Physical Review Letters*, 66:2056–2059, April 1991.

[15] W. Chen and Y.-S. Wu. Implications of a cosmological constant varying as R^{-2}. *Physical Review D*, 41:695–698, January 1990.

[16] C. B. Collins and S. W. Hawking. The rotation and distortion of the Universe. *Monthly Notices of the Royal Astronomical Society*, 162:307, 1973.

[17] F. Cooper, A. Khare, R. Musto and A. Wipf. Supersymmetry and the Dirac equation. *Annals of Physics*, 187:1–28, October 1988.

[18] P. A. M. Dirac. The Cosmological Constants. *Nature*, 139:323, February 1937.

[19] G. F. R. Ellis. *Cosmological models from a covariant viewpoint.*, volume 211 of *Astrophysics and Space Science Library*. Springer, Netherlands, Dordrecht, 1997.

[20] W. L. Freedman, B. F. Madore, V. Scowcroft, C. Burns, A. Monson, S. E. Persson, M. Seibert and J. Rigby. Carnegie Hubble Program: A Mid-infrared Calibration of the Hubble Constant. *Astrophysical Journal*, 758:24, October 2012.

https://doi.org/10.1515/9783110515367-202

[21] L. P. Grishchuk and Y. V. Sidorov. Squeezed quantum states of relic gravitons and primordial density fluctuations. *Physical Review D*, 42:3413–3421, November 1990.
[22] A. H. Guth. Inflationary universe: A possible solution to the horizon and flatness problems. *Physical Review D*, 23:347–356, January 1981.
[23] P. Havas. The range of application of the Lagrange formalism — I. *Il Nuovo Cimento*, 5:363–388, October 1957.
[24] K. Husimi. Miscellanea in Elementary Quantum Mechanics, II. *Progress of Theoretical Physics*, 9:381–402, April 1953.
[25] K. C. Jacobs. Spatially Homogeneous and Euclidean Cosmological Models with Shear. *Astrophysical Journal*, 153:661, August 1968.
[26] E. Kanai. On the Quantization of the Dissipative Systems. *Progress of Theoretical Physics*, 3:440–442, October 1948.
[27] D. C. Khandekar and S. V. Lawande. Exact solution of a time-dependent quantal harmonic oscillator with damping and a perturbative force. *Journal of Mathematical Physics*, 20:1870–1877, September 1979.
[28] S. P. Kim and C. H. Lee. Nonequilibrium quantum dynamics of second order phase transitions. *Physical Review D*, 62(12):125020, December 2000.
[29] N. A. Lemos and C. P. Natividade. Harmonic oscillator in expanding universes. *Nuovo Cimento B Serie*, 99:211–225, 1987.
[30] J. A. S. Lima and J. C. Carvalho. Dirac's cosmology with varying cosmological constant. *General Relativity and Gravitation*, 26:909–916, September 1994.
[31] A. D. Linde. A new inflationary universe scenario: A possible solution of the horizon, flatness, homogeneity, isotropy and primordial monopole problems. *Physics Letters B*, 108:389–393, February 1982.
[32] I. A. Malkin and V. I. Manko. Dynamic symmetry and coherent states of quantum systems. *Moscow Izdatelstvo Nauka*, 1979.
[33] H. G. Oh, H. R. Lee, T. F. George and C. I. Um. Exact wave functions and coherent states of a damped driven harmonic oscillator. *Physical Review A: General Physics*, 39:5515–5522, June 1989.
[34] T. S. Pereira, C. Pitrou and J.-P. Uzan. Theory of cosmological perturbations in an anisotropic universe. *Journal of Cosmology and Astroparticle Physics*, 9:6, September 2007.
[35] W. Rindler. Relativity: special, general, and cosmological. In *Relativity: special, general, and cosmological*. Oxford University Press, Oxford, UK, 2001.
[36] H. C. Rosu and V. Ibarra-Junquera. FRW barotropic zero modes: Dynamical systems observability. *ArXiv General Relativity and Quantum Cosmology e-prints*, April 2006.
[37] H. C. Rosu and P. Ojeda-May. Supersymmetry of FRW Barotropic Cosmologies. *International Journal of Theoretical Physics*, 45:1152–1157, June 2006.
[38] E. Russell, C. B. Kılınç and O. K. Pashaev. Bianchi I model: an alternative way to model the present-day Universe. *Monthly Notices of the Royal Astronomical Society*, 442:2331–2341, August 2014.
[39] B. Saha. Nonlinear spinor field in Bianchi type-I cosmology: Inflation, isotropization, and late time acceleration. *Physical Review D*, 74(12):124030, December 2006.
[40] B. Saha. Early inflation, isotropization, and late time acceleration in a Bianchi type-I universe. *Physics of Particles and Nuclei*, 40:656–673, September 2009.
[41] C. P. Singh, S. Kumar and A. Pradhan. Early viscous universe with variable gravitational and cosmological 'constants'. *Classical and Quantum Gravity*, 24:455–474, January 2007.
[42] S. Weinberg. *Gravitation and Cosmology: Principles and Applications of the General Theory of Relativity*. Wiley-VCH, July 1972.

[43] K.-H. Yeon, D.-H. Kim, C.-I. Um, T. F. George and L. N. Pandey. Relations of canonical and unitary transformations for a general time-dependent quadratic Hamiltonian system. *Physical Review A: General Physics*, 55:4023–4029, June 1997.

Index

https://doi.org/10.1515/9783110515367-203

De Gruyter Studies in Mathematical Physics

Volume 40
Joachim Schröter
Minkowski Space: The Spacetime of Special Relativity, 2017
ISBN 978-3-11-048457-1, e-ISBN (PDF) 978-3-11-048573-8,
e-ISBN (EPUB) 978-3-11-048461-8, Set-ISBN 978-3-11-048574-5

Volume 39
Vladimir K. Dobrev
Invariant Differential Operators: Quantum Groups, 2017
ISBN 978-3-11-043543-6, e-ISBN (PDF) 978-3-11-042770-7,
e-ISBN (EPUB) 978-3-11-042778-3, Set-ISBN 978-3-11-042771-4

Volume 38
Alexander N. Petrov, Sergei M. Kopeikin, Robert R. Lompay, Bayram Tekin
Metric theories of gravity: Perturbations and conservation laws, 2017
ISBN 978-3-11-035173-6, e-ISBN (PDF) 978-3-11-035178-1,
e-ISBN (EPUB) 978-3-11-038340-9, Set-ISBN 978-3-11-035179-8

Volume 37
Igor Olegovich Cherednikov, Frederik F. Van der Veken
Parton Densities in Quantum Chromodynamics: Gauge invariance, path-dependence and Wilson lines, 2016
ISBN 978-3-11-043939-7, e-ISBN (PDF) 978-3-11-043060-8,
e-ISBN (EPUB) 978-3-11-043068-4, Set-ISBN 978-3-11-043061-5

Volume 36
Alexander B. Borisov
Nonlinear Dynamics: Non-Integrable Systems and Chaotic Dynamics, 2016
ISBN 978-3-11-043938-0, e-ISBN (PDF) 978-3-11-043058-5,
e-ISBN (EPUB) 978-3-11-043067-7, Set-ISBN 978-3-11-043059-2

Volume 35
Vladimir K. Dobrev
Invariant Differential Operators: Volume 1 Noncompact Semisimple Lie Algebras and Groups, 2016
ISBN 978-3-11-043542-9, e-ISBN (PDF) 978-3-11-042764-6,
e-ISBN (EPUB) 978-3-11-042780-6, Set-ISBN 978-3-11-042765-3

Bei Fragen zur Produktsicherheit wenden Sie sich bitte an:
If you have any questions regarding product safety,
please contact:

Walter de Gruyter GmbH
Genthiner Straße 13
10785 Berlin
productsafety@degruyterbrill.com